Jossey-Bass Teacher

Jossey-Bass Teacher provides educators with practical knowledge and tools to create a positive and lifelong impact on student learning. We offer classroom-tested and research-based teaching resources for a variety of grade levels and subject areas. Whether you are an aspiring, new, or veteran teacher, we want to help you make every teaching day your best.

From ready-to-use classroom activities to the latest teaching framework, our value-packed books provide insightful, practical, and comprehensive materials on the topics that matter most to K–12 teachers. We hope to become your trusted source for the best ideas from the most experienced and respected experts in the field.

Supporting Mathematical Learning

Effective Instruction, Assessment, and Student Activities

Joanne Picone-Zocchia
Giselle O. Martin-Kniep

JOSSEY-BASS
A Wiley Imprint
www.josseybass.com

Published by Jossey-Bass
A Wiley Imprint
989 Market Street, San Francisco, CA 94103-1741—www.josseybass.com

Jossey-Bass books and products are available through most bookstores. To contact Jossey-Bass directly call our Customer Care Department within the U.S. at 800-956-7739, outside the U.S. at 317-572-3986, or fax 317-572-4002.

Jossey-Bass also publishes its books in a variety of electronic formats. Some content that appears in print may not be available in electronic books.

Library of Congress Cataloging-in-Publication Data

Picone-Zocchia, Joanne.
    Supporting mathematical learning : effective instruction, assessment, and student activities/ Joanne Picone-Zocchia and Giselle O. Martin-Kniep.
      p. cm.—(Jossey-Bass teacher)
    Includes bibliographical references and index.
    ISBN 978-0-7879-8876-0 (alk. paper)
    1. Mathematics—Study and teaching (Elementary)—Activity programs—United States. 2. Lesson planning—United States. 3. Effective teaching—United States. I. Martin-Kniep, Giselle O., 1956– II. Title.
    QA135.6.P536 2008
    372.7—dc22

Printed in the United States of America

FIRST EDITION

PB Printing   10 9 8 7 6 5 4 3 2 1

About the Book

Blending theory of "best practice" with classroom reality, this book focuses on the development and use of effective instructional techniques and practices in K–5 mathematics. Including concrete problems, lessons, and activities that are rigorous and engaging, it explores the topics of diversified assessment; establishing and communicating explicit criteria, using various questioning strategies to support and extend math learning; and the role of problems in providing context for teaching and learning math skills. Chapters comprised entirely of lessons provide examples of teacher- and student-centered learning experiences, as well as suggestions for differentiating lessons to meet students' diverse needs.

About the Authors

Joanne Picone-Zocchia, a teacher for twenty-two years before becoming a professional developer, has a background that includes elementary, secondary, and special education. She now serves as vice president of Learner-Centered Initiatives, Ltd., where she consults with schools and districts on the design and use of a standards-based curriculum, instruction and assessment, portfolio and rubric development, and practices that promote student thoughtfulness.

Picone-Zocchia is also the vice president of operations and organizational development for Communities for Learning: Leading Lasting Change®, an organization established to assist schools, districts, and other educational entities in developing professional learning communities that serve to build capacity and expertise within both individuals and organizations. This role involves her in a variety of facilitation efforts, from implementing systemic reform processes to working with teachers on identifying and implementing best practices in mathematics and other subjects.

Additionally, Ms. Picone-Zocchia facilitates "trainer of trainer" programs geared to supporting those who lead the learning of other adults. She also assists districts in the design and implementation of new-teacher induction programs and instructional practices that address the needs of diverse learners. She is a published curriculum writer and has co-authored work on the use of curriculum maps to assess teachers' work.

Giselle O. Martin-Kniep is a teacher educator, curriculum designer, researcher, program evaluator, and writer. She is the president of Learner-Centered Initiatives and the founder and president of Communities for Learning: Leading Lasting Change®.

Dr. Martin-Kniep has a strong background in organizational change, and has graduate degrees in communication and development, social sciences in education, and educational evaluation from Stanford University. She has taught at Adelphi University, the University of British Columbia, and the University of Victoria. She has worked with hundreds of schools and districts, nationally and internationally, in the areas of curriculum and assessment, standards-based design, school improvement, and action research.

Dr. Martin-Kniep has produced and edited numerous curriculum materials for the Stanford Program in International and Cross-Cultural Education (SPICE) and for Communities for Learning. Her books include *Why Am I Doing This: Purposeful Teaching with Portfolio Assessment,* published by Heinemann; *Becoming a Better Teacher: Eight Innovations that Work;* published by ASCD; and *Developing Learning Communities Through Teacher Expertise,* published by Corwin Press. Her most recent book is titled *Communities That Learn, Lead and Last: Building and Sustaining Educational Expertise;* published in 2007 by Jossey-Bass. She is now working on a new book centered on a framework for effective teaching, which will be published by ASCD.

Communities for Learning: Leading lasting change™ is a nonprofit organization established in 1997 and located in Floral Park, New York, whose mission is to create professional communities that learn, lead, and last. Originally named the Center for the Study of Expertise in Teaching and Learning (CSETL), this organization has worked to improve the learning of schools as organizations, as well as that of the adults and students inside them, through the identification, consolidation, and dissemination of educators' expertise and best practice in the context of professional learning communities. Operating as a professional learning community functioning at the systemic level, Communities for Learning supports fellowship programs in which teachers, administrators, professional developers, university faculties, and students reconcile their individual passions and expertise with the vision of the organizations that sponsor them.

Communities for Learning offers a variety of programs and services related to the development and support of professional learning communities. These include readiness assessments; skill-based programs; keynotes and workshops; programs for understanding change; and facilitation of school-based, network, and regional professional learning communities, as well as programs that lead to the certification of a school, district, or region as a Communities for Learning: Leading Lasting Change site and facilitator training and certification programs designed to support those individuals and organizations wishing to establish and sustain Communities for Learning sites.

For further information, contact:

Communities for Learning: Leading lasting change™
249-02 Jericho Turnpike, Suite 203
Floral Park, NY 11001
516-502-4232 phone
516-502-4233 fax
www.communitiesforlearning.org

Acknowledgments

Special thanks to Julianne DesOrmeaux for her time, sample problems, and feedback; Buffalo and New York City public schools, teachers, and students for their assistance in developing and field testing problems and lessons; Kathy Davis for her expertise in math for primary students; and to all of the Communities for Learning Fellows who made themselves available at a moment's notice for inspiration, feedback, or moral support.

This book is dedicated to all of the Communities for Learning which are leading lasting change in education, and to their participants, whose learning and work are instrumental in making this happen.

Contents

Part II: Lessons and Activities

Contents   **xiii**

Contents

Chapter 1

Introduction

Effective Mathematics Instruction

*A problem is not necessarily solved because
the correct answer has been made.
A problem is not truly solved unless the learner
understands what he has done and knows
why his actions were appropriate.*

—William A. Brownell,
The Measurement of Understanding (1946)

Teaching mathematics is a complex enterprise. Mathematics involves a language, a body of knowledge, a way of thinking, and a set of skills. Having mathematical proficiency involves acquiring several skills and dispositions. According to the National Research Council (2002), these include computing, applying, understanding, reasoning, and engaging in mathematics. Mathematical proficiency also involves mastery of different content strands. The National Council of Teachers of Mathematics Standards classifies these strands as numbers and operations, algebra, geometry, measurement, and data analysis and probability. Alongside the content strands are also the process strands of problem solving, reasoning and proof, communication, connections, and representation. Whereas there are multiple frameworks for classifying and organizing mathematics, there is little disagreement about the competencies and attitudes that effective mathematics thinkers and users display.

Students who know and use mathematics well are able to understand it. They can comprehend mathematical concepts, operations, and relations—knowing what mathematics symbols, diagrams, and procedures mean. They can also interpret the mathematical principles in a problem and translate those ideas into a coherent mathematical representation using the important facts of the problem.

Proficient students can compute well. They can carry out mathematical procedures—such as adding, subtracting, multiplying, and dividing numbers—in a way that is accurate, flexible, efficient, and appropriate. Computation and execution require that students learn skills so that they can remember them, apply them when needed, and adjust them to solve new problems.

Proficient students can formulate problems mathematically and can devise strategies for solving them using concepts and procedures appropriately. They can reason well by using logic to explain and justify a solution to a problem or to extend it from something known to something unknown.

Requiring students to present their reasoning processes and, where appropriate, to justify them, can help them reflect on their thinking, identify mistakes, and improve their strategies. It also enhances their communication and discussion skills, while enabling teachers to identify the emergence of powerful mathematical ideas among students, so that these can be clarified and nurtured. Reasoning and communication go hand in hand.

Effective mathematical communication requires that students see the connections and relationships in the things they know. It involves using the language of mathematics to express ideas as well as to organize and consolidate mathematical thinking through communication.

Proficient students develop mathematical insights. They can recognize the significance of a problem and its relationship to other problems, other disciplines, or "real world" applications. They understand how mathematics ideas connect and build on each other to produce a coherent whole.

Despite an apparent agreement on the knowledge and skills that mathematics requires, teachers' own background and experience often lead them to adopt a particular view of mathematics and an accompanying approach to teaching it. According to Ernst (1988), teachers tend to have one of the following three conceptions of mathematics:

1. *Problem-solving view:* mathematics as an expanding field of human creation and invention, in which patterns are generated and then distilled into knowledge. Mathematics is seen as a process of inquiry and coming to know, with its results open to revision.

2. *Platonic view:* mathematics as a static but unified body of knowledge, with interconnecting structures and truths. From this perspective, we discover mathematics rather than create it.

3. *Instrumentalist view:* mathematics as a bag of tools made up of an accumulation of facts, rules, and skills used for utilitarian purposes.

Teachers' beliefs and understanding of mathematics have a significant impact on students' beliefs about the nature of mathematics. Many teachers believe that mathematics is the discipline of the "right answer." This belief prevents them from helping students express their mathematical thinking, learn from their mistakes, experiment effectively, and pursue their mathematical interests. Teachers need to help transform the student question *Am I right?* into the questions *How can I develop confidence and judgment that I am on the right track when working on a problem?* and *How can I know that I am improving my mathematical problem-solving and communication skills?* (Math Forum Bridging Research and Practice Group).

Most students in the United States believe that doing mathematics requires a lot of practice in following rules. According to a 2003 study of K–12 mathematics education, only 15 percent of the K–12 mathematics lessons in the United States would be considered high in quality. The factors that distinguish effective lessons from ineffective ones are teachers' ability to do the following:

- Engage students with mathematics content by relating mathematics problems to the real world, connecting to students' interests, and promoting their active involvement.

- Create an environment conducive to learning, through the use of interesting and engaging problems whereby the teacher encourages multiple solution methods, supports students' conceptual understanding, and encourages mathematical thinking.

- Ensure access for all students by allowing them different entry points to problems that are challenging and that require persistence. Good problems are nonroutine, unfamiliar, and just beyond the student's skill level so that the student does not automatically know how to solve them.

- Use questioning to monitor and promote understanding through the diversified and strategic use of content and of processing, leading, nonleading, and clarifying questions.

- Help students make sense of the mathematics by using content that connects to other problems and mathematical concepts, aligning their teaching with current mathematics curriculum and standards and integrating mathematics with other subject areas.

Some Strategies for Improving Mathematics Instruction

To improve student engagement with mathematics content, teachers can develop or adapt problems so that these have a real-life context or purpose. They can also remind students of problem situations that are conceptually similar to the ones they are solving, and they can provide them with needed background knowledge on the problems they have to face. Another strategy for promoting engagement is to lead students through instant replays of a problem situation and encourage them to request the assistance of the teacher or other students.

Among the strategies that teachers can use to create a classroom environment that is conducive to learning is the use of problems that can generate many solutions. Teachers can maximize the use of such problems by waiting for, and listening to, students' descriptions of solution methods, as well as encouraging students to elaborate on their problem-solving strategies and solutions. When teachers use students' explanations as a basis for the lesson's content, they convey an attitude of acceptance of students' errors and efforts.

Mathematics instruction can be greatly enhanced through the use of diversified questioning skills. Such skills include knowing how to use leading, nonleading, clarifying, and processing questions that tap students' different levels of thinking. Leading questions can be used to begin a lesson, probe the depth of students' understanding, elicit content, help a student clarify or extend his or her thinking, or provide a focus; for example, *How did we get the length of this desk? Why is this a circle?*

Nonleading questions are most effective when responding to students' ideas. The context defines whether a question is leading or not. Teachers ask nonleading questions when we are trying to facilitate students' thinking; for example, *What happened? What did you observe? How did you get it? Why are you asking that question? Why does it work?* Chapter Four provides teachers with different strategies for enhancing their use of questions in mathematics.

An effective use of wait time can enhance teachers' questioning skills. Wait time includes providing time for students to find the words for an explanation and listening patiently as students try to put their questions into words. It is also a time for teachers to ponder options before acting. Much can happen as a result of acting as though we believe that students have much to say.

Teachers can improve their instruction in mathematics through the development of their communication skills. For example, paraphrasing students' answers to questions shows students that they are being listened to; it can also help introduce mathematical terms to students. Prompts that denote paraphrasing include: *What I hear you saying is . . .* or *Do you mean that . . . ?* Another effective communication skill involves summarizing. Summaries provide students with a compact record of their thinking that they can revisit later. Finally, listening is also part of the overall effort to diagnose students' strengths and needs in order to determine appropriate interventions. It involves listening to understand by recognizing that students' mathematical actions and explanations are reasonable from their point of view, even if the reason is not immediately apparent to us. (Cobb et al., 1991).

In many classrooms, students are expected neither to explain their thinking about mathematics nor to justify it. If students are taught and supported, they can present their reasoning processes clearly and can justify them as well. Even young students can present their thinking processes quite clearly once they are encouraged to do so and know that their thinking is valued. The use of explicit criteria in the form of rubrics and checklists as instructional devices, along with many of the process questions included in Chapter Three, will help students articulate and develop their reasoning skills.

The chapters in this book model the development and use of effective instruction. Chapter Two introduces teachers to diverse assessment measures and processes. Through the use of concrete problems and situations, teachers learn about diagnostic, formative, and summative assessment, and explore the uses of recall-based, performance, product, and process assessment. The chapter ends with an overview of portfolio assessment and the role it can play in the overall evaluation of students' growth and achievement.

Chapter Three introduces teachers to performance criteria related to different content and processes in mathematics and enables them to discover many different ways of articulating their expectations for student achievement. It also explores ways of involving students in the assessment process and helps teachers identify their own preferences with respect to rubrics and checklists.

Chapter Four introduces teachers to diverse questioning practices and helps them identify questions that tap different thinking and reasoning processes. Teachers learn about convergent and divergent questions, questions that tap different levels of Bloom's Taxonomy (Bloom, 1956), and the use of process questions to help students develop conceptual understanding, reasoning, and other mathematics competencies and dispositions.

Chapter Five showcases the use of problems as a means of contextualizing the teaching of mathematical skills. This chapter includes a number of annotated, open-ended problems so that teachers can see the possible range of students' approaches or answers to each of them. The primary intent of this chapter is to help teachers see the use and value of problems as learning and assessment opportunities.

Chapters Six through Nine highlight the use of engaging mathematics lessons centered around patterns, measurement, money, and fractions, many of which embed problems like the ones included in Chapter Five. These chapters are structured so that teachers can compare teacher-directed lessons with student-centered lessons that address the same content and skills. These lessons include explicit differentiation strategies as well as formal assessment opportunities.

The appendices include an annotated list of Web-based and other resources for teaching and assessing in mathematics.

Teaching Elements

Chapter 2

Assessment

Measuring Student Learning

To *assess* is to determine what students know, think, and are able to do.

"A high quality assessment system relies on a variety of assessments to provide timely and understandable information to all who need it, so that they can make the instructional decisions that maximize student success" (Chappuis, Stiggins, Arter, & Chappuis, 2004, p. 28).

In classroom assessment, the connections between teaching and assessing enable and promote student success. The tighter and more explicit these connections are, the more useful the assessment information will be. *Alignment, purpose and timing*, and *assessment type* are three key links in the relationship among assessment, teaching, and learning. Attending to and strengthening these links strengthens the classroom assessment system as a whole, providing a flow of current and relevant data that can inform ongoing instruction and improve learning.

Alignment

The first critical link in the assessment-teaching-learning relationship is the tight alignment between what is taught and what is assessed. A mismatch between the two results in assessment information that has little or no connection to the original learning target and is of questionable use in terms of promoting student achievement. The concepts, processes, content understandings, and skills assessed should match those that were taught. Establishing clear outcomes and measurable indicators is an important step in strengthening this link.

Outcomes are broad, global statements that describe what students will ultimately know and be able to do. Most state standards are stated in way that are equivalent to outcomes. As these are interdisciplinary in nature, it can be difficult to determine which subject area an outcome describes, especially if read out of context. Because of their breadth and global nature, outcomes must be supported by indicators, which are more specific statements that define the measurable steps students need to take to reach an outcome. These indicators provide the actual assessable components of outcomes.

An example of an outcome for math is: *Students will solve a variety of problems.* To determine indicators, use a stated outcome followed by the word *by*, as though you were engaging in a sentence completion activity. Think about your curriculum and expectations, and complete the outcome sentence in as many ways as make sense, being sure to identify measurable steps and not activities. Here's an example of this process using the outcome *Students will solve a variety of problems:*

Students will solve a variety of problems by:

- Engaging in a variety of problem-solving activities
- Applying problem-solving strategies appropriately
- Analyzing problems for important information in order to understand what must be done
- Organizing information
- Explaining steps taken or strategies used
- Finding multiple ways of solving a problem
- Using math language correctly

Examples of steps that are *not* indicators include "solving word problems using guess and check," "creating problems for other students to solve," and "writing explanations of their problem-solving process in their math journals." Rather than measurable steps, these describe the specific classroom tasks or activities that provide opportunities or contexts for one or more of the actual indicators to be assessed.

Outcomes and their indicators provide the framework within which all teaching and learning is focused. When designing assessments, the indicators that were taught are the indicators that should be measured. It is critical that assessments measure the learning that is likely to stem from the learning opportunities actually provided in lessons and units.

Purpose and Timing

The second key link in strengthening the connections among teaching, learning, and assessment is a focus on the purpose and timing of the assessments implemented. *Why am I assessing this?* and *Why am I assessing this right now?* are two important questions to guide assessment design.

Strategic classroom assessment systems incorporate assessments that attend to issues of purpose and timing. Gathering information about what students know before they engage with the lessons that are planned, monitoring learning while students are developing understandings and skills, and evaluating student achievement at the culmination of a lesson set or unit—each of these represents a unique and important opportunity to assess student needs and make related instructional decisions. Called, respectively, *diagnostic, formative,* and *summative* assessment, together these assessment opportunities provide educators with a rich and multifaceted understanding of student learning.

Discovering What Students Already Know

Diagnostic assessment, a form of assessment that occurs before teaching or learning takes place, responds to these questions. It is a low-stakes assessment for students, in that there are no related grades, rewards, or consequences. Information from the diagnostic assessment is used to plan subsequent learning experiences.

The purpose of diagnostic assessment is to help the teacher learn about the content understanding, skills, and relevant experiences that students bring to a learning experience. Tapping into students' prior knowledge, misconceptions, or misunderstandings before beginning a learning experience helps teachers to plan instruction based on reality rather than on what they assume or suspect. Diagnostic assessments can also become part of conversations about student placement or services.

Diagnostic assessment can take many forms. Examples of the kinds of activities that can be used as diagnostic assessment include semantic or concept maps, surveys, and KWL charts (charts that ask students to identify what they Know about something, what they Want to know about that topic, and what they Learned as a result of a lesson or unit). Tests or quizzes can also be used, as long as the teacher's purpose remains to gather information about what students already know and are able to do related to the topic, and to use that information in planning the lessons and assessments that will follow. Sometimes diagnostic tests or quizzes are called *pre-tests*. In this case, the same or similar questions can be administered on completion of the unit or learning experience, allowing both teachers and students to measure growth and learning.

In math, diagnostic assessment can be helpful in identifying students' abilities with respect to specific, necessary content, skills, or concepts. Presenting problem-solving situations as diagnostic assessment in math provides teachers with information about a broad range of student attitudes, skills, and knowledge. What follows is a variety of examples of diagnostic assessments.

Diagnostic Assessment Example 1

Primary Problem

This problem provides information about students' abilities to sort, categorize, and explain their reasoning. Responses can vary. For example, some students will group by shape, by color, or even by straight sides as opposed to curves. Explanations will vary as well and should be assessed based on the degree to which they explain the relationship of the items in the groups.

Provide small groups of students with a set of the following shapes, made from construction paper:

Squares: 1 red, 1 blue, 1 yellow
Circles: 1 yellow, 1 red
Triangles: 2 blue, 1 red
Rectangle: 1 yellow

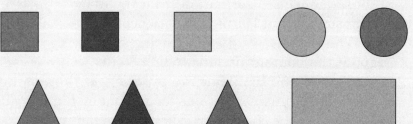

1. Ask students to separate the shapes into groups so that every shape in the group is the same as every other shape in the group.

2. As they form their groups, have students identify a rule for each.
3. Prompt them to find as many different ways as they can.

Diagnostic Assessment Example 2

Intermediate Problem

This problem is appropriate for assessing the degree to which students can engage with word problems. It will provide data on how students unpack a problem and whether or not they can isolate the information they need to solve it. In addition, teachers can assess a student's ability to multiply or divide.

The Problem:
Toni is throwing a surprise birthday party for her younger brother and needs to buy party supplies. She is expecting thirty people to attend. She plans on buying small bags of M&Ms and single bags of potato chips. She wants to give each person the same amount of candy and chips. The bags of M&Ms come in boxes of six, and the potato chip bags come in boxes of five. How many boxes of chips and how many boxes of M&Ms does she need to buy?

Diagnostic Assessment Example 3

Numbers with Zero

Diagnostic question: When you write all the numbers from 1 to 99, how many times do you write a zero?

Student response:

10 20 30 40 50 60 70 80 90
1 2 3 4 5 6 7 8 9
I rote the tens to ninety and then I numberd them. My answer is 9.

Follow-up question and answer:

Q – Why didn't you write all numbers from 1–99?

A – It would take too much time. I only wrote the numbers that have zeros.

Diagnostic Assessment Example 4

Math Word Map Diagnostic

This diagnostic assessment helps teachers to see what students know about math vocabulary and also allows for a degree of integration of personal experience and knowledge with the definition. This diagnostic assessment can be extended into an ongoing learning experience by having students periodically revisit and update their maps.

(Cont'd.)

1. Discuss with students what it means to learn new words, and review the parts of a definition.
2. Introduce and describe the math word map organizer. Model with a familiar math word.
3. Complete the word map together for several new words from the unit or lesson.
4. Allow students to independently complete word maps.

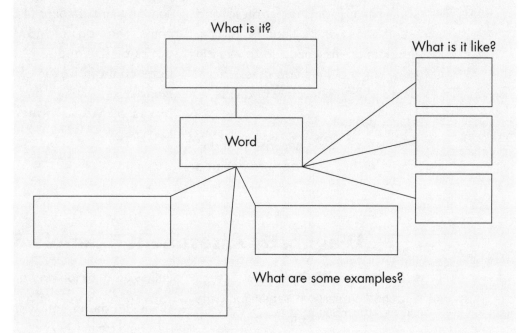

What is it?

What is it like?

Word

What are some examples?

Math Word Map

A student has completed the following diagnostic word map on the term *graphs*. Notice the student's understanding of the need for two axes as well as the phrases written along the connecting lines, which complete some of the relationships between and among words.

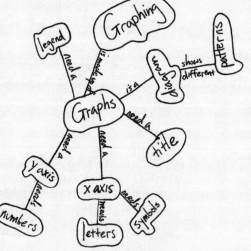

Monitoring What Students Are Actually Learning

Formative assessment is an opportunity to peek inside learners' heads while they are in the process of learning. Its purpose is to provide ongoing information about how and what students are learning and to create opportunities for feedback to the learner. Occurring throughout teaching and learning, formative assessment is about finding moments to check students' understanding, measure where students are with respect to what is being taught, and make needed instructional adjustments based on this information.

As with diagnostic assessment, formative assessment is a low-stakes assessment for students. Its power to impact learning, however, is tremendous, due to the feedback that it supports (for teachers, about the relative effectiveness of lessons and activities; for learners, about the degree to which their learning and work are meeting expectations and standards) and the revisions that it inspires. Formative assessment helps teachers and learners alike answer the question, *What do I do next?* Teachers use information from formative assessments in planning interventions when appropriate and adjusting subsequent learning experiences; students use it to help set goals, make revisions, and alter approaches or choices. Most of the assessment that occurs in a classroom is, or should be, formative.

In mathematics, formative assessment can help teachers to monitor students' developing skills and understanding. This can be done in a variety of ways. Math journals, small group review of problem sets, pairing students to share results and give feedback, asking students to explain their process or student-teacher conferences are all examples of formative assessment measures appropriate to the math classroom.

The use of problems as formative assessment opportunities incorporates much of the formative data provided by the preceding list into one activity that allows teachers to gather a broad range of information, from reading to computation to reasoning and communication. The process of problem solving lends itself to formative assessment opportunities. During problem-based learning experiences, students can have opportunities for self-, peer-, and teacher assessment, giving and receiving feedback, and making revisions. Conferences around problem-solving strategies and their implementation, reflective journal entries about process and results, as well as an explanation of the actual steps taken in finding the solution to the problem, are also appropriate formative assessment opportunities related to problem solving.

The following examples of formative assessment are annotated to indicate what information the teacher could gather about student learning and what, if any, interventions might be appropriate. The journal prompts were created by math teachers from MS 45 (NYC, Region 1) to support the IMPACT Mathematics program (Glencoe/McGraw-Hill). The first prompt includes a sample student response, annotated to indicate formative information and suggested next steps for the teacher.

Journal Prompt

Describe what a perimeter is and how to find it for various shapes. Give an example of a situation in which finding a shape's perimeter would be useful.

Student Response:

Perimeter is the line around the outside of something. To find the perimeter, you would measure the different sides and add them together. That would tell you how far the perimeter is. If you wanted to put a rug in a room, you would need to know the perimeter so you could tell how big the rug could be.

Analyzing Formative Assessment Information

In this example, the student evidences an understanding of the definition of *perimeter*. There may be some question about the application of that definition, however, because the student responds to the part of the prompt that asks how to find the perimeter of various shapes by giving a generic, though correct, explanation. It is not clear whether or not the student can differentiate, for instance, between determining the perimeter of a circle and that of a polygon. Additionally, the example provided of when finding a shape's perimeter would be most useful raises questions about whether the student is actually thinking about the fit of the outside edge of the rug to the walls of the room, or whether the student is confusing perimeter with area.

Based on the questions raised as a result of reading this student's response, the teacher might decide to meet with the student for clarification and to obtain a more complete response. If misunderstandings do exist, the teacher can then address them specifically.

More Sample Prompts

- If one shape has a greater area than another, must it also have a greater perimeter? Explain or illustrate your answer.
- What have you learned about different uses for graphs?
- Create your own graph and explain what it shows.
- Make a graph that shows how something changes over time. Write a story to go with your graph.

Evaluating What Students Have Learned

Summative assessment occurs at the end of a learning experience, at a time when it is important to evaluate the degree to which students have actually learned what was taught. It can be, among other things, the test at the end of a unit, a project, or a research paper. This is a high-stakes assessment for students, because it is tied to evaluation of their learning. It is summative assessment that impacts grades, promotion, placement, graduation, and so on.

In math, there can be several kinds of summative assessment. Frequently tests are used, but assessments such as presentations, research papers, projects, or a combination can also provide summative assessment information. As is the case in diagnostic and formative assessment, incorporating problem-solving experiences into summative assessment is an efficient way of evaluating multiple levels of student learning within a single experience. Well-designed problems, as well as variations of problems that have been used for diagnostic and formative purposes, can enable teachers to gather data on students' abilities to compute, reason, apply concepts and skills, communicate, and make connections.

Assessment Types

Yet another key link important to a strong relationship among assessment, teaching, and learning is the *type of assessment* implemented. From *information recall* to *product, performance,* and *process,* using different types of assessments provides a degree of flexibility and versatility that allows many layers of learning to be measured, and a degree of precision that permits specific information about student learning to be documented.

For example, if the desired data relates to what students can remember of what was taught, then an *information recall* assessment (also called a *recall* assessment) makes the most sense, so brief, content-based questions are appropriate. Recall assessments focus on what students can remember from what they have learned. They can be completed individually or collaboratively. Recall assessments are objective, including items such as multiple choice, true or false, matching, fill-in, or short-answer questions.

To know what students can create as a result of what they have learned requires *product* assessment. Product assessments can be completed individually or collaboratively and result in a tangible product such as a poster, research report, essay, sculpture, or lab report. Product assessment can be fully or partially authentic (that is, having a real purpose and audience outside of school), depending on the degree to which the product created has a real audience and a purpose or context that is bigger than the assessment experience itself.

Product assessment is one form of measuring students' abilities to apply what they have learned, but it is not the only way. If a teacher wants to know what students can do as a result of what they've learned, then designing a *performance* assessment is best. Performance assessments can be completed individually or collaboratively, involving students in performing in some way. Performance assessments can take the form of a role-play, demonstration, oral or panel presentation, skit, or playing a game. As with product assessment, performance-based assessment can be fully or partially authentic.

Though each of the preceding types of assessment has an important role in measuring student learning, none of the types mentioned so far would be particularly helpful in determining how students are thinking about their learning—where they recognize that they may be having difficulty or what insights they may have developed as the result of a lesson or unit. For such purposes, a *process* assessment is called for; this could take the form of a metacognitive activity or a series of reflective prompts. Process assessments focus on how students think, learn, work, write, or problem solve. Process assessments encourage students to share their thoughts about their learning through the use of journals, verbal reflection, logs, and "think alouds."

Recall, product, performance, and process assessment specific to math can be further broken down into three subcategories: skills tasks, problems, and projects (Balanced Assessment in Mathematics Program, 2005).

1. *Skills tasks:* tasks that primarily test the ability to manipulate and compute. *For example:* At a party, twenty bags of candy were given out. Each bag contained five candies. How many candies were given out altogether?

2. *Problems:* tasks that primarily test the ability to model, infer, and generalize. *For example:* Which two rooms in your school are farthest apart? Figure out three different routes that go from one to the other and tell how you would decide which is the shortest route.

3. *Projects:* multiday tasks that test the ability to analyze, organize, and manage complexity. *For example:* How many types of advertisements have you seen? Do they have anything in common? Are there certain types of items that are advertised more than others? Collect as many different ads as you can from magazines or newspapers. Identify a list of categories so each of your ads belongs to one of those categories (if someone had an ad for baseball bats,

for example, you might want a category titled "sporting goods"). Construct a chart or graph to show how many ads you found of each type. Explain which category seems to be more popular with advertisers.

Awareness of different types of assessment allows teachers to be strategic in structuring their assessments to best fit their intended purpose, helping them to accurately measure student learning. The ability to purposefully tailor an assessment activity to the kind of assessment information that is desired is a powerful skill. Balancing a classroom assessment system so that it includes opportunities for students to engage in each type of assessment allows students to show what they know in a variety of ways, acknowledging and honoring the presence of diverse learning styles and intelligences.

Evidence of Learning: Assessment Data

Ascertaining student learning is no small task. Learning can occur in subtle or overt ways and can seem minimal or monumental. This is further complicated by the varying degrees of readiness and prior experience or knowledge that students bring to learning experiences, leaving teachers with the often-difficult task of trying to determine what actually constitutes learning in a given situation.

Using Pre/Post Assessment

How can we assert that we know precisely what a student has learned, let alone that it was learned as a result of what was taught? Implementing pre/post assessment can simplify the process of claiming that learning has occurred by defining the parameters of the learning event, providing the evidence of what was before and what is now, and uncovering the degree or depth of learning as a result of comparing the two.

A pre-test can be used as a diagnostic assessment, allowing the teacher to assess what each student brings to the learning experience in the way of prior knowledge and understanding, gaps in skills or experience, or misunderstanding. Used as a diagnostic assessment, a pre-test can allow the teacher to craft learning experiences that connect to, and build on, this reality. A pre-test can also become the baseline measure for each student, against which other work can be compared to determine change, growth, and learning.

A post-test is a form of summative assessment whose purpose is to show what students know at the end of a lesson, activity, or unit.

When compared with the results of the pre-test, the post-test can show evidence of change or growth in skills or understanding—the learning that occurred as a direct result of the learning opportunities presented. Without the pre- and post-tests, it is difficult to determine what learning actually resulted from the lessons or activities presented, and what learning happened around them—or even in spite of them. Implementing pre/post assessment allows these connections to be made.

Sample Pre/Post Assessment

The following sample uses an essential question as a pre/post assessment. In addressing the question, "Are numbers real?" students provide evidence of their conceptual understanding, mathematical reasoning, and abstract thinking. This also provides students with an opportunity to make connections between numbers and real life. In the following student example, there were approximately two months between the pre and the post.

Pre-Test Response:

Sure numbers are real. If I have three pens, that's real. And if I have thirty dollars, that's really real. You can write numbers and use them to measure and count. All of that is real.

Post-Test Response – Journal entry:

I think that numbers are real, but sometimes they are more real than other times. If numbers are used to count things or to measure size or distance, then they are pretty real because they are connected to something that is real. But if numbers get really big or really small, or if they are negative, then they just mean something in your head, and I think that makes them less real. Like infinity. Is infinity real? How do we know? Or 0.00000000000000000000002. What is that? Is that real?

Another thing that makes numbers real or not is if you understand them. Numbers are real when they mean something to the people using them. XVXII meant something to Ancient Romans, so it was real to them. It doesn't mean anything to me, so I don't think it is real. Scientists use scientific notation, but since I'm not a scientist, a number like $34^{23671345}$ isn't real to me.

When we did our Marketing Unit, we had to order candles and figure out how much money we would make and how much we could spend buying more candles to sell. Sometimes, we were just thinking about numbers that could be—like how many candles we would probably sell the next day and how much money we could invest in new candles. So I don't think those numbers were really real, because they were guesses. But the numbers that the accountants reported every day were very real, because there was money in the lock box that matched those numbers. The number of candles that were sold was real because we didn't have them anymore. And the number of candles that we bought was real because they came wrapped in boxes and we had to pay for each one. The numbers that told us our profit at the end were real, too.

I think numbers are real when we use them to mean things that we have or that we understand.

In another classroom, the teacher asks students to put on a piece of paper everything they know about fractions. They are prompted to write, draw, chart, create a graphic organizer—whatever will help them show what they know or believe about fractions. These are collected by the teacher and used to indicate students' understanding or misperceptions about fractions. The information helps the teacher design lessons that meet the specific needs of this group of students.

After the unit on fractions, the teacher gives the students a new piece of paper with the same instructions as before: "Use this paper to show all that you know about fractions. You may draw, write, chart, create a graphic organizer—anything that will help you to show what you know or believe about fractions." The teacher can assess this post-test and compare it with the pre-test to determine what individual students have learned. Students can be given their pre- and post-tests and asked to identify what they know now that they didn't know before.

Not only does this provide assessment data, but also, when students look at their own pre- and post- assessments, they compare themselves to themselves and recognize their own accomplishments, without seeing them in the context of the other students in the class. This can be especially powerful for low-achieving students, allowing them to see and document learning and growth.

Portfolio Assessment in Mathematics

"A portfolio is a collection of student work that exhibits the student's efforts, progress, and achievements in one or more areas. The collection is guided by a clearly established purpose and has a specific audience in mind" (Martin-Kniep, 2001, p. 66).

Student-centered portfolios maximize students' critical thinking and provide engaging and deep insights into students' learning, beliefs, and perceptions of themselves as learners. They support students as strategic learners and promote a sense of student ownership of, and responsibility for, their learning that is difficult to find inside other assessment structures.

When students produce work that is then graded and returned or filed, that work and its worth exist in the moment. Although its importance may once again rise to the surface at the end of a grading period or year, its significance as an event is primarily relegated to the moments in which it is produced by the student and assessed by the teacher.

When students sift through the body of work that they have produced in order to identify the strongest example or examples possible as evidence indicating achievement of, or progress toward, a curriculum goal or a learning outcome, it changes their perspective about their learning and their work. Work whose moment had passed, which would likely have had no reason to surface again, suddenly becomes critical as potential evidence of their learning. Outcomes and curriculum goals, formerly the purview of the teacher, now explicitly exist within the realm of the student, defining the essence of what they are responsible for both achieving and proving that they have achieved. The need to clearly connect their work as evidence of their learning adds a degree of relevance and meaningfulness to the work itself, and situates the responsibility of proving achievement, improvement, and learning as one in which the student shares. Establishing a student-centered portfolio process as part of the classroom assessment system provides the structure within which this shift in perspective and responsibility can take place.

What Kind of Portfolio Is Best?   Teachers can select from several kinds of portfolios, each of which serves a different primary purpose:

- *Showcase portfolios* document achievement or potential. They generally include samples of what students consider to be their best work.
- *Development portfolios* show growth or change over time. They include evidence of the process by which work is done as well as the final product, and they often trace the evolution of one or more projects or products.
- The purpose of a *transfer portfolio* is to communicate with students' subsequent teacher, school, or employer.
- A *keepsake portfolio* is a collection of favorite or memorable work. Combinations of any of these types of portfolios are common and often allow for a more comprehensive view of the student as a learner.

Developing classroom math portfolios provides an opportunity for both teachers and students to focus carefully and thoughtfully on teaching, assessment, and learning related to specific and important outcomes in mathematics.

Developing a Classroom Portfolio Assessment System   In developing a portfolio assessment system, the first thing to do is to decide on a primary purpose and audience for your students' portfolios. If you cannot make a decision between two primary audiences or purposes, then develop two different portfolios. Some teachers like the idea of having students create a school portfolio and a home portfolio. Others create nested portfolios; for example, a showcase portfolio of best work might exist inside a development portfolio.

The next consideration is to determine whether the portfolio will be outcome-driven or curriculum-content-driven. If outcome-driven, the portfolio content will evidence student achievement, effort, or growth toward certain student-learning outcomes. If curriculum-driven, the portfolio's purpose will document student understanding of specific curriculum content.

Once this determination has been made, the next step is to identify the learning outcomes and the specific indicators that the portfolio will assess (determined for a portfolio using exactly the same process as for a unit or course) or the curriculum areas that the portfolio entries will seek to make evident.

Finally, it is necessary to create the prompts, or directions, that will enable students to select work that evidences their learning related to the outcomes or curriculum, and to reflect on their choices. These selections and accompanying reflections are the *portfolio entries*.

Sample Portfolio Entries
- Letter to the reader explaining the portfolio
- Table of contents
- Periodic goal statements and strategies for meeting goals
- Excerpts from the student's journal
- Drafts and final versions of student's work
- Self- and peer evaluations
- Teacher- and student-completed checklists
- Homework and assignments
- Sample of most challenging work
- Work that shows corrections of errors and misconceptions on assignments or tests
- Report on a group project, with commentaries on the individual's contributions

A broad range of entries is acceptable, including but not limited to journal entries, computer CDs, designs or inventions, reports, response logs or reviews, posters or other artistic media, collaborative works, attitude surveys, problems, projects, self-assessment checklists, teachers' comments and checklists, parental comments, and test scores and tests.

The Portfolio Selection Prompt   A selection prompt is a teacher-generated question or statement that guides students as they select pieces, or entries, for their portfolio. Selection prompts should be aligned with the portfolio outcomes. Selection prompts can be used with students at various times throughout the year, as well as at the end of the year.

Here are some sample prompts teachers use to guide students' selections for their portfolios:

- Select one sample that shows you have improved as a problem solver.
- Select one sample that shows you have worked toward the goal set in the first quarter.
- Select one sample that shows that you are persistent.
- Select one sample in which you use a graph, table, chart, or diagram to inform others.
- Select one assignment that you could have changed to improve it, and describe such a change.
- Include a problem that took more than one class to solve; explain the steps you went through to complete it and why you decided on those steps. Include notes, drafts, and anything else that would give me an idea of what you had to do.
- Include a problem that you solved with at least one other person in class. Describe how each group member worked together and their contributions to this work.
- Include work that has corrections you made to errors, additions, or other changes. Explain why they were necessary.

In PS/MS 315, in the Bronx, K–8 teachers decided to create the following three school-wide portfolio outcomes, which would apply to all subject areas:

- Students will write effectively for a variety of purposes and audiences.
- Students will become problem solvers.
- Students will become reflective learners.

The following are two examples of the organizer that the teachers used to align their outcomes, indicators, learning opportunities, and prompts.

K–8 Outcome 1: Students will write effectively for a variety of purposes and audiences.

Content Area	Indicators:	Learning Opportunities:	Prompts:
	(What are the measurable steps* that I can see students will take or achieve as they work toward this outcome in my class?) *Could be more or fewer than three.	(What lessons, activities, assessments, and so on do I have planned that will provide students with opportunities to accomplish these indicators?)	(What statements or questions can I pose to my students that will help them to select work that provides evidence of their learning, progress, or achievement related to the outcome or indicator?)
in Math …	by Reflecting on learning at the end of a lesson. Explaining how they got an answer. Using math vocabulary correctly.	Reflect every other day on the lesson and what they've learned. Respond to reflections during and after assessment. Discuss strategies used to solve problems. Write in math journal. The math message.	Put something in your portfolio that shows you know how to explain how you got an answer. Put something in your portfolio that shows you understand, and can use, a math strategy. Put something in your portfolio that shows you can use math vocabulary correctly.

K–8 Outcome 3: Students will become reflective learners.

Content Area	Indicators: (What are the measurable steps* that I can see students will take or achieve as they work toward this outcome in my class?) *Could be more or fewer than three.	Learning Opportunities: (What lessons, activities, assessments, and so on do I have planned that will provide students with opportunities to accomplish these indicators?)	Prompts: (What statements or questions can I pose to my students that will help them to select work that provides evidence of their learning, progress, or achievement related to the outcome or indicator?)
in Math …	by Responding to reflection questions, participating in reflective activities that require them to think and write about themselves as math learners. Responding to reflection questions, participating in reflective activities that require them to self-assess and comment on their math learning, process, or a product that they have created. Setting and monitoring goals related to self-assessment based on specific and explicit expectations and criteria—their own as well as their teacher's.	Formal written responses to reflection questions related to extended, multi-step word problems. Written reflections at the end of class. Written reflection related to opportunities to use criteria in self- and peer-assessment feedback. Formal opportunities to set and monitor goals based on feedback or on assessment, using criteria.	Choose one or more reflective pieces to put into your portfolio that show your growth as a math thinker and learner. Put something in your portfolio that shows that you can set, and work toward, specific and measurable goals that you have set for yourself as a result of self-, peer, or teacher feedback.

One of the things that quickly became apparent was how important the indicators would be, as those would make the work appropriate for each grade-level cluster. Additionally, it didn't take long to recognize that students would need help to understand some of the outcomes in concrete ways. Mini-lessons were set up to help students get a sense of what was important to consider for each pick. The following is some of the work of a fourth grade class related to understanding problem solving in multiple contexts.

What Does Problem Solving Look Like?

Reading	Writing	Math	Science	Social Studies
Use of context clues Prediction Visualizing Inferring Rereading Connections Partnerships	Persuasive letters Sharing ideas with people Editing Revising	Math boxes Problem of the day Algebra Fact families Math messages Operations Math warm-up	We make hypotheses as to what might happen. We discover different information. Question Experiment	How Native Americans helped us—they helped us with food. Past is connected to the future.

Question: What does it mean to be a problem solver?

Student responses:

A person that makes a problem gone

Solves problems like a scientist

When there is a problem there is someone to solve it

Helps you to solve problems like Ms. Eileen

There is a way to fix a problem

Like a detective

Solves problems—makes things easier to understand

After engaging in the mini-lesson, students were prompted, "Select something for your portfolio that shows you can write about what you have learned in math." What follows is a sample cover page for a portfolio pick, with scaffolding to help students focus on responding to the activity.

I have chosen _____ to show how well I can write about what I have learned in math. When you read it, you will see that I have learned how to _____

_____

_____

_____

You can tell this because I write_____

_____

_____

_____

Teacher Reflection:

Maintaining a portfolio affects everyone's role and perspective in the classroom. Teachers move from being dispensers of information and knowledge to becoming designers of learning experiences whose purpose is to provide students with multiple opportunities to access the portfolio's outcomes and indicators. Students shift from being recipients of information and developers of skills to becoming identifiers of evidence of learning and accomplishment. Even the role of student work changes, from being products of the curriculum to becoming evidence of student learning measured against specific outcomes and indicators.

Assessment as a System, Not an Event

Thinking about assessment as a series of interactive opportunities, all of which together build a comprehensive and vibrant image of student learning, provides the evidence necessary to support evaluation while preserving the integrity of the connections among teaching, learning, and assessing—none of which can happen well or responsibly without the others in place. An effective classroom assessment system is one in which assessment opportunities clearly measure the intended outcomes of lessons, activities, and units, and are strategically timed and designed to maximize their effect on both teaching and learning. It allows students to self- and peer assess, giving, receiving, and incorporating their own and peer feedback, as well as teacher feedback and ultimate evaluation. Attending to the alignment, purpose and timing, and types of assessments used, and assessing students' understanding of taught concepts, content, processes, and skills in a variety of ways throughout the learning experience, provides an optimum opportunity to garner information and evidence relative to what is being learned and how.

Chapter 3

Setting and Communicating Explicit Criteria

The term *criteria* refers to quality attributes of work or performance. These communicate expectations and support the selection of models. Whether diagnostic, formative, or summative; self, peer, or teacher; whenever assessment is the goal, criteria are at the center of the picture. This chapter discusses the development and use of *explicit criteria*—that is, clear and precise language to communicate quality attributes and expectations—so that teachers can access and apply them consistently, and students can understand and use them to inform and support their learning and their work.

Having access to explicit criteria and structured opportunities to use them during the course of a unit, a series of lessons, or an activity or project can help students assume greater responsibility for their work and demystify the assessment process. Finally, the use of explicit criteria is necessary if we want to document students' progress toward, and attainment of, city, state, and national standards. Teachers can share their criteria for quality in a variety of ways, including verbally, using models or exemplars, and by developing checklists or rubrics. Each of these choices has its own distinct advantages and limitations. Although all four strategies will be discussed, this chapter will focus primarily on the development of checklists and rubrics and their use as instructional and assessment tools.

Verbal Criteria

One of the most common methods of communicating criteria is doing so verbally. Explaining what is expected can be highly interactive, with the advantages of engaging students in conversation and clarification in the moment. Verbal sharing of criteria requires no prep time beyond the time it takes to think through what the expectations actually are. It is simple and direct.

The limitations of verbal criteria are directly related to its advantages. The very nature of this kind of sharing requires and presumes a degree of attending and aural processing that may or may not actually exist. Distractions of any kind can remove student attention from the moment and result in their missing the criteria, entirely or in part. Because it is difficult to capture conversation such that it can be revisited at a later time, without careful documentation, explanations and agreements are left to the respective memories of those who were in the room. Those who are not present miss out entirely.

Models and Exemplars

The use of models or exemplars is another way of sharing criteria. The adage, "A picture is worth a thousand words," supports this strategy. Using models or exemplars provides concrete examples for students of what expectations actually look like in practice. These allow students, especially those who have difficulty imagining an end product, to refer to images, rather than descriptions of quality, and they are particularly helpful for visual learners.

However, models and exemplars are of limited use without time to explore the ways in which they embody the quality criteria they are illustrating. A second limitation lies in the number of models or exemplars presented. Using one or two can result in students perceiving that there are limited ways of achieving quality, and this may result in copies of the examples rather than original work. Using three models or exemplars that show a range of possibilities tends to diminish this effect. With primary and elementary students, using more than three exemplars is probably unnecessary and could cause some confusion among the students.

Checklists

A third way for teachers to communicate quality criteria is through the use of a checklist—a list of components or expectations for a learning experience, activity, or assessment. The list can be organized to support student learning and work by clustering the criteria into labeled categories, prioritizing the checklist in terms of importance, or listing criteria in an order that supports students' completion of the task. In the following checklist example the criteria are clustered and in an order that supports students' completion of the task.

Checklist for Presentation of Survey Data

Data

☐ Data is factual and relevant to the topic.

☐ Data reflects student survey results.

Bar Graph

☐ Bar graph format is accurate, legible, and neat.

☐ Vertical and horizontal axes are properly labeled and identified.

☐ Bar graph reflects the data collected.

☐ Bar graph illustrates the survey.

As a written form of communication, a checklist has a degree of stability that allows its criteria to be revisited over time, enabling students' and teachers' repeated use during the course of a learning experience or assessment.

A variation on the checklist is a point scale, which lists the components of a task or process with associated points or ranges of points. It is possible to be awarded anywhere from none of the available points to all of them for each of the components. Although this clarifies, to some extent, issues such as relative importance of components and provides some sense of quality, it puts the emphasis on the score and stops short of explaining the specifics that are required to achieve the most points.

A checklist is simple, clear, and easy to navigate, using few words. It quickly allows both students and teachers to establish the presence or absence of each expectation or component. Yet this distinct advantage can also be the checklist's greatest limitation: its very brevity can make it seem cryptic to some students, and terminology that is not completely familiar can require considerable unpacking to be understood well enough to be helpful.

The checklist, useful in ascertaining whether or not an attribute is present, has little or no connection to how well the attribute may be executed. Because of this, it is impossible for teachers or students to use a checklist to support revisions aimed at improving quality. For that, one needs a rubric.

Rubrics

A rubric, yet another way for teachers to share their criteria, is a scale that defines and differentiates levels of performance. Unlike a checklist, which determines the presence or absence of specific attributes or components, a rubric presumes their presence and works to assess quality. A rubric can be analytic or holistic, and it can describe the qualities attached to generic processes or specific tasks, as well as levels along a developmental continuum.

An analytic rubric is designed to identify domains (dimensions) of the task and describe each of its attributes (descriptors) separately. These

Dimensions	4	3	2	1
Identifies coins (penny, nickel, dime, and quarter) by their name.	You can name all the coins taught and other money not taught.	You can name all of the coins taught.	You confuse the names of the coins taught.	You still need to learn the names of the coins taught.
Knows the value of the penny, nickel, dime, and quarter.	You know the value of all of the coins taught, as well as other money not taught.	You know the value of all the coins taught.	You know the values of the coins taught, but you confuse them.	You need to know the value of the coins taught.
Counts money using pennies, nickels, dimes, and quarters up to $1.50.	You can count money amounts beyond $1.50, using all the coins on your own.	You can count amounts of money to $1.50, but you need to be reminded to use all the coins.	You can count amounts of money, but you don't use all the coins, even when reminded.	You can count amounts of money using only the penny.

rubrics have limited descriptors for each dimension at each level of achievement. The example of an analytic rubric on page 34 describes four levels of quality related to the identification and value of money. In this example, each dimension has only one descriptor, but it is possible for dimensions to have multiple descriptors as well.

Analytic rubrics allow for specific assessment information. Their specificity makes them ideal for providing feedback that students can use in self- or peer assessment, as well as to revise their work. Analytic rubrics also provide scaffolding for the learning experience, helping students to identify specific actions that they can take or changes they can make to improve their performance, product, or process. Using the previous example, students not only would know at what level they were currently performing but also could tell by looking at the next level up what they needed to do to improve.

Another type of rubric is a holistic rubric. Like analytic rubrics, these can be generic, developmental, or task specific. Holistic rubrics are designed to capture the big picture—the entirety of a product. They rely on multiple descriptors for each level of achievement, but do not identify separate dimensions in which those descriptors cluster. An example of a holistic rubric is below.

Holistic rubrics are limited in value in terms of providing precise assessment information. They are also limited with respect to the specificity of feedback that students can use them to give or receive, and the

Recycling Project Rubric

How you practiced math and followed the directions for building

3 – Gives a complete and detailed response and a clear explanation. Writes schedule correctly, with events following one another and including times. Makes a list with dollars and cents for each piece used and adds it all up without a calculator. Writes sentences about the graph that include correct math.

2 – Gives an incomplete response and explanation. Includes a schedule with events following one another. Makes a list with dollars and cents for each piece used. Graphs the shapes correctly.

1 – Gives an incorrect response and explanation. Lists events without apparent order. Includes a list of recyclables used and counts the shapes.

degree to which the rubric in and of itself can scaffold a learning experience or help students revise their work.

Developing Rubrics and Checklists

Although it is always a good idea to use the reality of student work to inform checklist and rubric design in some way, circumstances do not always allow for student work to be the impetus behind the initial design, as would be the case when using a newly designed assessment. Without student work samples, both checklists and rubrics are grounded in one of the following kinds of information:

- The task itself
- Related teacher expectations
- Expectations from state or other standards

Designing a Checklist

Being mindful that the purpose of a checklist is to provide students with a tool for documenting the presence of specific, important components of a product, performance, or process, one begins developing a checklist by brainstorming the component parts of a task or assessment. Checklists can be written in the first person ("My solution ..."), second person ("Your solution ...") or the third person ("The solution to the problem ...").

Step 1: Identify and list all the attributes or indicators that make a quality performance, process, or product. Think about both content and form; for example:

- Explanation includes math vocabulary.
- All computations are shown.
- Work has been checked.
- Work is neat.
- Work is organized.
- Math vocabulary is used correctly.
- Explanation is clear.

Step 2: Look for performance indicators in the standards documents that relate to the product, process, or performance but that are not on your list. Add them. For example: "Develop and use strategies to estimate the results of whole-number computations and to judge the reasonableness of such results." *Note:* If student work is available, look at several examples to see if you have captured everything you want about the criteria for the work.

Step 3: Cluster these attributes or indicators in possible groups or categories and label each group (dimension) with a heading. For example, the indicators, "Explanation uses math vocabulary," "Math vocabulary is used correctly," and "Explanation is clear" can be clustered under *Communication.* Write a definition for that heading in the form of a question (such as "How well did you explain your work?") or a statement (such as "The degree to which you explained your work").

Step 4: Prioritize each group in terms of importance or value or determine the order that will be most useful for your students as they work.

Designing a Rubric

Although a checklist is about the presence of specific components, a rubric goes a step beyond, describing the quality of those components at different levels of achievement. Developing a rubric begins in much the same way as developing a checklist, with the articulation of the important components and qualities of the task at hand. These are then clustered into categories and the categories are titled and defined. The components in each category are then described at different levels of achievement.

Checklists and rubrics can work well together, with the checklist assessing the presence of components prior to quality assessment, or the checklist assessing components whose presence or absence is their most important attribute. In this case, the checklist would inform the level of expectation of the rubric. This would be the third level in a four-point scale.

Step 1: Using the checklist you developed, select either the most important dimension/cluster or the one that students must deal with first. For that cluster, write descriptions for the following levels (in the order suggested):

Level 3 (created first): Develop the third level of your rubric by describing the behaviors, characteristics, or qualities that exhibit a proficient product, process, or performance. The language from your checklist will be the basis for the proficient level of your rubric. For example, "You use math vocabulary correctly" and "Your explanation is clear" are descriptors for Level 3.

Level 4: Describe the behaviors, characteristics, or qualities that exhibit achievement at the highest level. The highest level should be above the expected standard of excellence. For example, "You use math vocabulary precisely" and "Your explanation is clear and gives reasons for the decisions you made" are descriptors of Level 4.

Level 1: Write the descriptors for the lowest level of the rubric. Identify what the student has done, not just what the student has not done; for example, "You use math vocabulary incorrectly" and "Your explanation is misleading or has errors."

Level 2: Write the descriptors for the second level of the rubric. Identify the nature of the errors students make at this level; for example, "You use math vocabulary imprecisely" and "Your explanation shows confusion or misunderstanding."

Step 2: Select another dimension and write the descriptions for each of the levels. Complete the remaining dimensions.

Step 3: Use the rubric (multiple days).

1. Share the rubric with students, engaging in conversations around clarity.
2. Revise based on confusions or questions.
3. Have students use the rubric while developing their performance, product, or process.
4. Incorporate the rubric into multiple formative assessment moments, allowing rubric-based feedback to lead to time to revise and improve work based on the criteria presented.
5. Use the rubric as a summative assessment tool.

Step 4: Revise the rubric for clarity and precision, using student work generated from the task.

Involving Students in Developing Explicit Criteria

Involving students in the process of developing criteria and designing rubrics or checklists can help them relate to and understand the criteria. When students are involved in the process, they

- Think about quality
- Internalize much of what is on the rubric
- Are able to understand the language of the rubric
- Are able to use the rubric more easily
- Feel ownership of the rubric
- See the rubric as an instructional tool that they can use
- See the rubric as an assessment tool that they can use
- Become partners in the assessment process

There are many ways to involve students in establishing and describing criteria. Students can

- Identify attributes of quality for a product or process
- Cluster attributes
- Identify titles of the clusters (the dimensions of the rubric)
- Define the dimensions
- Draft a rubric or checklist, in whole or in part
- Think about and discuss the importance and weighting of attributes
- Give feedback on the strengths and weaknesses of a rubric
- Refine a rubric to make it more useful

Using Criteria to Support Classroom Assessment

Rubrics have come to be recognized as an assessment tool, but their greatest power actually lies in their use as a teaching and learning tool. Once the rubric or checklist is designed, it can be used to support classroom assessment in a variety of ways. The frequency and variety of experiences that include its use will greatly influence the degree to which students internalize the criteria it expresses. Providing multiple opportunities for students to use the rubric or checklist with their own and each other's work is crucial in making the connection between criteria and learning.

To best enable the connection between criteria and learning, teachers can

- Provide multiple, formal opportunities for students to use rubrics and checklists as formative assessment (self- and peer assessment, goal setting, revision guide)
- Tie rubrics and checklists to performances and processes that are important to both the teacher and the students
- Break up a rubric and target a specific dimension or dimensions to support mini-lessons, individualized goals for students, or a lesson focus
- Establish that all feedback, whether student or teacher, revolves around the criteria articulated in the rubric or checklist

To best use the criteria expressed in rubrics and checklists, students can use a rubric or checklist

- To self-assess a product or performance
- To guide reflection about process or achievement
- For peer assessment

- When used for self- or peer assessment, as a basis in setting goals for future work
- To monitor the student's own progress or learning

Clearly and specifically establishing and expressing criteria enables both teachers and their students to better understand the expectations of any given learning experience. With that understanding as a base, opportunities to apply explicit criteria in the form of rubrics and checklists provide the steps necessary for better attainment of these expectations and help students develop as independent and responsible learners.

Chapter 4

Using Questions Effectively

*Good questions recognize the wide possibilities
of thought and are built around varying forms of
thinking. They are directed toward learning and
evaluative thinking rather than determining what
has been learned in a narrow sense.*

—N. M. Sanders,
Classroom Questions: What Kinds?

Questioning is a natural form of human discourse and a powerful instructional tool. Questioning is the second most popular teaching method (after lecturing). In fact, teachers spend 35 to 50 percent of their teaching time on questions.

Teachers ask students questions for many different reasons. In the early grades, a chief reason is to develop students' social skill of raising and responding to questions. Across all grades, teachers use questions to develop students' skills, to conduct inquiry and ask questions, to increase the level of student participation, and to increase higher-order thinking skills. Teachers also use questions to ascertain the effectiveness of their lessons and to manage their pace and direction. In terms of classroom climate, teachers use questions to control student behavior, empower students, embarrass them, or remind them of who is in control of the classroom. In this chapter, we will explore the use of questions as a means to enhance and produce learning in mathematics.

A review of the research on questioning by the Northwestern Research Educational Laboratory and other studies reveal a number of questioning practices that promote learning (Cotton, 2001; Walsh & Sattes, 2003; Walsh & Sattes, 2005). Such practices include incorporating questions into lessons (Gall, et al., 1978) and asking such questions *during* lessons rather than posing them before or after lessons.

The frequency and timing of the questions teachers ask have an effect on students' acquisition and internalization of content. Students do better on test items previously asked as questions. In general, learning of facts is enhanced when teachers ask questions frequently during discussions, but learning of complex material is not enhanced with more frequent questions. Regardless of their frequency, oral questions posed during teacher presentations are more effective than written questions. Very young children and poor readers need advance questions that help them look for information they will generate later. Older children benefit from more open-ended questions before reading or studying (Cotton, 2001; Sitko & Slemon, 1982; Swift & Gooding, 1983).

The types of questions teachers ask can make a significant difference in student achievement, and not all questions are equivalent in terms of supporting students' learning. For example, questions that focus on salient aspects of lessons are more effective than extraneous ones (Samson, Strykowski, Weinstein, & Wahlberg, 1987).

Questions can be classified in many different ways. These classifications can help teachers develop or broaden a repertoire of strategies that support different types of thinking and reasoning processes. For example, *memory* questions are used to assess students' recall of information. *Text-explicit* questions focus on information that can be extracted from a text or word problem. *Text-implicit* questions require that students "read between the lines." *Inference* questions center on implied meaning. *Interpretation* questions require that students restate or translate information into their own words. *Transfer* questions enable students to apply learning in one context or situation to another. *Predictive* questions ask students to ponder and state the relationship between two variables. *Evaluative thinking* questions require that students make judgments. *Reflective* questions focus on helping students think about their thinking and learning. *Investigation* questions require that students conduct an experiment. *Decision-making* questions demand that students make a decision, whereas *problem-solving* questions entail that they solve a problem. *Argumentation* questions demand that students take a position on an issue. *Classification or categorization* questions ask students to sort information, whereas *comparison questions* require that students compare and contrast data. The following table illustrates all these types of questions, with examples of mathematics questions.

Types of Questions	Example
Memory question: Provides background on a subject. Learners find the answers in sources such as texts, Web sites, and other reference materials.	What is the formula for the perimeter of a rectangle?
Text-explicit question: Requires little transformation of textual information.	How many students are going to the field trip?
Text-implicit question: Requires inferences or a combination of what the "reader" and the author have to say.	Which room will require more paint?
Inference question: Asks learners to go beyond the immediately available information.	What can you tell by looking at this chart?
Interpretation question: Demands that learners articulate the consequence of information and ideas.	How would the sum change if one addend increased by two?
Transfer question: Asks learners to take their knowledge and understanding to new contexts.	How would you use this technique to solve a money problem?
Question about hypothesis: Predictive questions.	What are your predictions about what may happen if we change the sizes of the pizza pieces?
Evaluative question: Deals with matters of judgment, value, and choice.	How completely and clearly did the student explain his or her process for solving the problem?
Reflective question: Asks students to think about their thinking and their learning.	What is getting in your way in solving that problem?
Investigation question: Demands the implementation of an experiment or lab.	How could we test which is the best strategy for this problem?

(Cont'd.)

Types of Questions	Example
Decision-making question: Asks students to make a decision.	What operation should we use?
Problem-solving question: Asks students to go through the problem-solving processes.	What is the solution to x?
Argumentation question: Asks students to take a position on an issue.	In this case, is it more important to be fast or to be correct?
Classification/categorization question: Asks learners to sort information into different categories or types.	In how many different ways can we sort these shapes?
Comparison question: Asks students to compare and contrast information.	How are these triangles different from and similar to these others?

Questions to Support Different Forms of Thinking

Teachers can also use questions that support the development and use of different forms of thinking. For example, questions can also be classified into *convergent* and *divergent* questions.

Convergent questions represent the analysis and integration of given or remembered information. They focus on a correct response. The questions can be factual and simple—*Where is the blue circle?*—or factual and complex—*What is the volume of your locker in cubic inches (not including the top box)?* In general, convergent questions are at the lower levels of Bloom's Taxonomy; they tap the processes of explaining, stating relationships, and comparing and contrasting.

Divergent questions are more demanding of a student's thought processes and may call for several plausible or correct responses. They often ask for students' opinions or conjectures. They tend to be at the upper levels of Bloom's Taxonomy and tap the processes of predicting, hypothesizing, inferring, or reconstructing. Sample divergent questions are *How do triangles influence space? Can numbers lie?*

The questions we ask can also be classified using Bloom's Taxonomy in terms of their tapping basic or higher-order thinking skills. Current research supports the use of higher-order questions to increase students' learning (Cotton, 2001; Walsh & Sattes, 2005). In fact, teaching all students to draw inferences results in increased cognitive responses and learning gains.

Following is a description of the different levels of Bloom's Taxonomy, the verbs associated with them, and sample questions for each level.

• *Knowledge* questions focus on helping students remember previously learned materials and recall appropriate information. These questions help students build an informational base for subsequent questions. The verbs commonly associated with knowledge level questions and prompts include *define, enumerate, identify, label, list, match, name, read, reproduce, restate, select, state, view, remember, memorize, recognize,* and *recall.* Sample questions are *What shape is this? How many brothers do you have? What is the best method for calculating the perimeter of a room? How many minutes are in a basketball game?*

• *Comprehension* questions require that students process information so that the meaning is clear. Comprehension is broken down into the subcategories of translation, interpretation, and extrapolation. The verbs and verb phrases commonly associated with this level of thinking questions include *classify, cite, convert, describe, discuss, estimate, explain, extrapolate, generalize, give examples, paraphrase, summarize, understand, interpret, translate, organize,* and *select facts and ideas.* Sample questions are *Given the amount of time we watch television today as compared to our parents, will television viewing by our own children increase or decrease? What are three ways you can read this clock's time?*

• *Application* questions help students use learned material in new and concrete situations. Application occurs in two phases. In the first, some abstraction, formula, equation, or algorithm is learned; in the second, students encounter a new situation or problem and are asked to apply it to a previous learning situation. Verbs associated with this level of questions include *act, administer, articulate, chart, collect, compute, construct, contribute, control, determine, develop, discover, establish, extend, implement, include, inform, instruct, participate, predict, prepare, preserve, produce, project, provide, record, relate, report, show, solve, teach, transfer, use,* and *utilize;* verb phrases include *apply information; produce results;* and *use facts, rules, and principles.* Sample questions are *How is _____ related to _____? Why is _____ significant? What formula should we use to find the length of this diagonal line?*

• *Analysis* questions promote the breaking down of material into component parts so that its organizational structure may be understood. In this level, students learn to take apart some complex phenomenon to show how it works. Words commonly used in questions at this level include *correlate, diagram, differentiate, discriminate, distinguish, focus, illustrate, infer, limit, outline, point out, prioritize, recognize, separate,* and *subdivide.* Sample questions are *How would you classify _____ according to _____?*

How would you outline/diagram _____? *How does* _____ *compare or contrast with* _____? *What evidence can you list for* _____?

- *Synthesis* questions require that students put parts together to form a whole. This level focuses on creativity and is product oriented. Words associated with questions at this level include *adapt, anticipate, categorize, collaborate, combine, communicate, compare, compile, compose, contract, contrast, create, design, devise, express, facilitate, formulate, generate, incorporate, initiate, integrate, intervene, model, modify, negotiate, plan, progress, rearrange, reconstruct, reinforce, reorganize, revise,* and *validate.* Sample questions are *What would you predict or infer from* _____? *What ideas can you add to* _____? *What might happen if you combine* _____? *What solutions would you suggest for* _____?

- *Evaluation* questions involve judging the value of material for a given purpose. Evaluation occurs in two steps: first, the establishment of some criteria, and second, the application of those criteria to some object or idea. Verbs and verb phrases commonly used with this type of question include *appraise, conclude, confront, criticize, critique, decide, defend, interpret, judge, justify, reframe, translate, make value decisions, resolve controversies,* and *develop opinions and judgments.* Sample questions are *What do you think about* _____? *What is the most important* _____? *How would you place the following in the order of priority:* _____? *How would you decide* _____? *What criteria would you use to assess* _____?

On average, approximately 60 to 95 percent of the questions most teachers ask are lower-order cognitive questions that revolve around factual information that can be memorized, 3 to 20 percent are higher-order cognitive questions requiring that students use their reasoning skills and skills to mentally manipulate information that has been previously learned to create or support an answer, and 5 to 20 percent focus on procedures. Students perceived by teachers as slow or poor learners are asked fewer higher-order cognitive questions than students perceived as more capable learners (Gall, 1984).

Several studies indicate that increasing the use of higher-order cognitive questions (above 20 percent) produces more gains for students above the primary grades, particularly for secondary students (Gall, 1984). Students attain higher levels of thinking when they are encouraged to generate critical and creative questions and when they can use these with their peers. In general, using a combination of higher- and lower-order questions is effective, but the balance between the two types of questions depends on the specific needs of students.

Teachers can use questions and prompts to help students develop key mathematical skills and traits, including conceptual understanding, strategies and reasoning, communication, computation and execution, and mathematical insights.

To help students develop their conceptual understanding through the interpretation of a problem's information, teachers can ask questions such as these:

- What is this problem about? Rewrite the problem in your own words.
- What do you know?
- What is the problem asking you to find or to do?
- What are the important facts and numbers in the problem?
- Is some information unnecessary for solving the problem?
- What math terms can help you understand and solve the problem?
- What will the answer look like (units of measure, level of accuracy required, form of the answer)?

To assist the development of students' understanding of a problem's mathematical concepts, teachers can ask questions such as the following:

- What types of computation will be required to solve this problem?
- Can you represent the problem so that it is easier to understand?
- What mathematical ideas and skills could help you represent and solve the problem (for example, graphing, identifying patterns, adding fractions)?

To help students develop their reasoning skills, teachers can ask the following questions:

- Can you work backward from where you want to end up to where you want to begin?
- Would drawing a picture or diagram or making a model be helpful in solving this problem?
- Would it be helpful to organize your information in a list, chart, or table?
- Would guessing, checking, and adjusting help you solve this type of problem?
- Should you be looking for patterns in your information?
- Would it be helpful to change the problem using simple numbers?

Questions that can help students think about their solution include the following:

- Are there other ways to approach the problem that might work?
- Can you give examples to support your solution?

- Is the strategy you used efficient?
- Is there a more efficient strategy to solve the problem?
- Can you explain your plan or strategy to someone else?
- Can you apply what you learned with other problems to solve this one?

Prompts and questions that help students communicate their thinking include the following:

- Can you use tables, graphs, pictures, words, or a combination of these in explaining or expanding your thinking?
- What did you do first? Why?
- What did you do next? Why?
- What (exactly) are you doing? Can you explain it?
- Why are you doing it?
- How does it help you?
- How did what you did help you reach the solution or goal?
- How did you figure out _____?
- What did you learn from doing this problem?
- How do you know that your answer is correct?
- Where can I find the process you used to solve this problem?

To help students improve their computational skills, teachers can ask questions like these:

- Did you double-check your calculations as you went?
- Did you estimate your answers before using a calculator?
- Did you show the rule or formula you used?
- Did you verify that your answer was correct by solving the problem in a different way or by plugging your answer into the problem to check if it made sense?
- Did you check to make sure your answer fits what the problem was asking for?

Prompts and questions to help students improve their mathematical insights include the following:

- Is your solution the only one that will work for this problem?
- How is this problem similar to or different from other problems you have done in class?

- How is this problem similar to other problems or situations you have seen in real life?
- Did you discover any patterns while you were solving this problem?
- What assumptions did you make in solving this problem?
- Can you create a problem that is like this problem in some ways but different in others?
- Can you find a process (or formula) that could be used to solve all forms of this problem?

Teachers can also help students develop their meta-cognitive skills through the use of reflective prompts and questions. Among the questions teachers can ask students before engaging with problems and tasks are the following:

- Do you think this problem will be easy or hard for you? Why do you think so?
- Do you have difficulty understanding any part of the problem? Describe or explain what you do not understand.
- Does the problem have any facts or information that aren't needed?
- Can you draw a diagram to illustrate the problem so that it is easier for me to solve it?
- What strategy do you think will help to solve the problem?

After students have completed a task or activity, teachers can ask questions such as these:

- Does your answer make sense according to the given facts?
- What strategy did you use? Why did you choose that strategy?
- Can you trace back your steps?
- Do you think your solution is correct? Explain why or why not.
- Was this problem easy or hard for you? Explain why.
- Did you follow a certain pattern?
- Can the same pattern be applied to other problems?
- Have you seen or solved a problem like this before? Describe such a problem.
- Could you have solved the problem in another way? Tell how without solving the problem again.
- Can you show me how to solve the problem?

Encouraging Student Questions

In addition to broadening the repertoire of questions teachers can use, teachers can support students' learning and achievement by helping them become better questioners themselves. To do this, teachers need to think of teaching questioning by exposing students to different types of questions, helping them classify and categorize them, modeling their use in different situations, and giving students ample opportunities to frame, ask, and answer their own and their peers' questions. The following are selected strategies to begin helping students ask better and different questions.

The first strategy, "Signal Words," consists of having students practice using the questions: Who? What? When? Where? Why? and How? To begin using this strategy, teachers can model the use of some or all of these signal word questions with a specific word problem. Here are some possible prompts: *Who* has a different way of solving this? *What* did she do? *When* should we use this operation? *Where* have we seen an example like this? *Why* did the number get smaller? *How* can we do this another way?

Another strategy for helping students ask questions is the adaptation and use of story or problem-solving structures. Such structures might include context or situation, variables, steps required for solving the problem, needed operations, and desired goal. Possible questions or prompts that can be used include *What is the problem here? What is it asking us to figure out? What do we know that is important? What is not important or relevant? What are the different things we have to do? What comes first? Second? What operations will we need to use? What will be determined when we complete all operations?*

Teachers can help students become better questioners by encouraging them to formulate and ask different kinds of questions. These may include many of the types of questions we have reviewed in this chapter, such as questions that tap different levels of Bloom's Taxonomy; questions that tap different thinking and reasoning processes; divergent and convergent questions; and questions that require different kinds of knowledge and connections from the learner.

To begin this process, teachers can engage students in a sorting activity with a common object, such as shoes, using the following steps

1. Students place all their shoes in a pile.
2. The teacher then asks students to identify all the different ways to sort the shoes. Students might do this by color, shape, size, newness, and material.

3. The teacher asks students to formulate questions about the shoes, such as, *How many shoes are _____? How old are _____? Where did the shoes come from? When is it better to use these shoes? Why do these shoes bend? Who made these shoes? Where does the material for the shoes come from? How much do these shoes cost? Who made the shoes? How long does it take to make shoes? What machines are involved in making shoes? How long do shoes last? What makes some shoes last longer? How many different sizes or shoes are there? What are the most popular shoes? Why do the shoes have different numbers inside? What part of the shoe breaks down more often? What happens to shoes after we throw them away?*

4. After students run out of questions to ask about the shoes, the teacher asks them to cluster the questions by their similarities and to give each cluster of questions a name. Possible cluster names are *why?* questions, *what?* questions, questions that can be answered with one word, questions that require an explanation, questions that can be answered on the spot, questions that require additional information, easy or five-second questions, harder or one-minute questions, and hardest or ten-minute questions.

5. The teacher can create and post this chart paper with the cluster names and sample questions in the classroom. This chart can become a question wall of types of questions that grows during the course of the year.

6. To encourage students to ask questions, the teacher can ask students to refer to the question wall and use it to formulate review questions, test questions, or discussion questions for the class.

Students' questions reveal at least as much about their thinking as their answers. Whereas their answers can confirm or disconfirm assumptions about what they understand, their questions reveal connections, misconceptions, interests, and insights. Teachers have a significant impact on students' learning, and the questioning practices they use have the potential of stimulating and enhancing students' learning.

Chapter 5

Quality Learning Problems

Skills in Context

It is a relatively common and comfortable practice to use problems and problem solving as a form of assessment, a way of determining the degree to which students can take skills and concepts they have learned and apply them to situations. In this case, *problem solving* is often made up of a series of word problems, having definitive right answers and generally requiring students to decode and transfer words to numbers.

Although it is crucial for students to understand concepts and learn skills in math, without context these concepts and skills hold little meaning or importance. Skills or concepts learned and practiced in isolation have little relevance beyond the practice sheet or classroom, and the degree to which they will be remembered or transferred to other situations is greatly compromised by the narrow frame within which they were initially taught.

Problems can begin to provide a context within which skills and concepts are learned for a purpose and practiced with intent. Using problems in this way means rethinking the role of the problem in the greater scheme of math education.

Although this can sound radical, there are a variety of approaches for using problems as a means to contextualize math.

For problems to begin to assume the much bigger role of providing context for mathematical learning, they must be open ended. "Open-ended questions are contextualized, interesting problem situations involving one or more mathematical concepts requiring student responses that include at least two of the following components: words, numbers and pictures or diagrams" (Dyer & Moynihan, 2000). Open-ended questions share certain characteristics; namely, they

- Have multiple entry points, allowing students of all abilities to engage with them and enabling teachers to observe and understand students' responses along a conceptual and procedural continuum
- Have multiple solutions and solution paths, or a single solution with multiple paths
- Focus on essential curriculum concepts in mathematics
- Involve student decision making
- Generate a need for students to communicate
- Tap real-world situations
- Allow students to work together to pose solutions
- Foster higher-order thinking
- Promote curiosity and pondering

At one extreme end of the spectrum is a problem-solving approach to math instruction and assessment, whereby problems are used to create a need for the content taught. To accomplish this, the problem is presented first, before students have all of the skills necessary to solve it. The needs presented by the problem are what determine both the content taught and its order. At the opposite end of the continuum are closed-ended problems used as summative assessments. These are the previously mentioned word problems, with definite correct answers, whose primary purpose is to test students' ability to transfer verbal information to numeric expressions and determine an answer.

More toward the middle of the spectrum, in this shift toward using problems to provide context, is a problem that supports a particular set of skills, concepts, or both. In this case, the open-ended problem is introduced just after the skills, concepts, or both are introduced and practiced. The problem provides an immediate application opportunity for targeted skills and concepts, and it gives students a sense of the importance of learning and using them.

The problems presented in this chapter are open ended and can be used in a variety of ways. Any of them could provide diagnostic, formative, or summative assessment information. All of them offer the potential of providing context for skills and concepts taught. Most of the problems that follow have been annotated to help teachers think about them in the context of the kinds of responses that students might give.

Grades K/1

Sample Problem 1: The Human Graph

This problem is embedded within a classroom activity. Engagement will be both physical and mental, as students become the actual pieces of data in a graph about their birthdays. The prompts given can provide students with insights about ordering relationships, as described in the annotations that follow.

Problem Guide:

Preparation: Post twelve signs on a single large wall, one sign for each month of the year.

Instructions to Students:

1. Stand in a line in front of the sign for your birthday month.
2. Which month has the most birthdays? How do you know? What's another way you could know?
3. Find somebody whose birthday is before your birthday. How did you decide?

After participating in this activity, students use Worksheet 5.1 and prepare individual graphs with round stickers to represent class members, labeling their own position in the overall graph. The creation of this graph allows students to cognitively move from the concrete level to one that is more abstract. To extend this problem, the preceding questions can also be asked about students' individual graphs. Students may compare graphs to those of others in their table group, or to the whole class.

Possible Responses:

- A typical response will be one in which the child creates the graph, possibly with assistance, but does not answer the questions in relation to either the human graph or the one he or she has constructed.
- A more sophisticated response is one in which the student can point to evidence on the graph created with stickers, showing that the connection has been made between the activity and its model.

Name _____   Date _____

My Birthday Chart

Jan	Feb	Mar	Apr	May	Jun	Jul	Aug	Sep	Oct	Nov	Dec

Sample Problem 2:
What's the Pattern, What's the Rule?

This problem provides a window into a child's thinking and learning style. Students are asked to provide the next picture or model and then state a pattern rule that would allow someone to continue creating pictures.

Here's the problem:

Look at the pictures in this pattern:

What would the next picture look like? Build it. Draw it.

How do you know it's the next picture?

In words, make up a rule for the pattern so that someone could continue making pictures that fit the pattern.

Although students at this age have some difficulty articulating a pattern rule, they generally have little trouble in extending the pattern. Responses, however, may show quite a large variation. Indeed, students occasionally provide a rule that is unexpected; it may still be correct if it meets the criteria for a pattern rule.

To be correct, the same pattern rule must explain the change from step 1 to step 2, from step 2 to step 3, from step 3 to step 4, and so on. In this problem, a pattern rule needs two components:

1. How many blocks should be added in the next step
2. The location of the added block(s)

In addition, the rule should be easily generalized to any step in the sequence. "If you have two blocks, make it be three blocks," is not in general terms, whereas "Add one block" can be generalized to any step in the sequence.

(Cont'd.)

Quality Learning Problems: Skills in Context

Possible Responses:

- A typical student response will be more like the first of the two preceding rule suggestions, and may not indicate the location of the new block without prompting from the teacher. Although most students will have no trouble building the next configuration, the articulation of a general rule will be difficult.

- A more sophisticated response is one in which the student independently generalizes either the number of blocks or the orientation of blocks, or both.

When the unusual response arises, it is important to listen carefully to the student's explanations and ask the same prompting questions that would be asked when viewing the expected solution. A configuration that appears at first to be quite random may be an alternate solution that is absolutely correct, just atypical.

From the processing of this problem, the teacher may see that some students actually prefer not to build the structure but to look at the problem entirely from a computational point of view, by counting. Others may refer to the number of blocks not as a quantity, but rather as relative quantities ("It gets higher, then stays the same"). Still others may look at the pattern as "building a box." In each case, the teacher is given insight into how individual children prefer to approach a problem in the manner that is most meaningful to them.

Grades K/1

Sample Problem 3: Making 16 Cents

Simple addition is required for this problem, in which computation is purposefully put in context. The use of manipulatives provides a means to add numbers that result in a sum of 16.

Several strategies may be employed by students when solving this problem. Some may be able to count by fives when using nickels; others may need to physically circle five pennies before replacing them with a single nickel (or ten for a dime). Each student (or group) has a sheet with images of coins, scissors, glue, and a blank sheet to record solutions.

The Problem:
You have several pages of dimes, pennies, quarters, and nickels. How many different ways can you find that will make 16 cents? Each time you find a way that works, cut out the coins and paste them into a box on your sheet with the blank boxes.

Possible Responses:

- A typical solution is one in which students are observed counting "by ones," even when using coins other than pennies. The student who is at the most concrete stages of development may need to physically replace five pennies with a nickel, for example.
- A more sophisticated solution is one in which students show evidence of counting by fives, or beginning with the dime as "ten." These students do not need to reference the "sixteen pennies" solution, but may reference the previous solution by replacing one dime with two nickels, or vice versa.

To extend this task for your students who show readiness, ask them to write the math sentences that show the addition they are using.

For example, a student who has used one dime, one nickel, and one penny may write:

$$10 + 5 + 1 = 16$$

or two sentences:

$$10 + 5 = 15 \text{ and } 15 + 1 = 16$$

Although a student's ability in computation will be indicated by the pasted coins on the worksheet, relating the problem to conceptual understanding and reasoning can be accomplished only through the child's response to various questions:

- Can you make 16 cents using only two coins? How do you know?
- What way uses the most number of coins?
- What way uses the least number of coins?
- Is there one kind of coin that you have to use?
- Is there one kind of coin that you can't use?
- Have you figured out all the possible ways? How do you know?

Quality Learning Problems: Skills in Context   **59**

Sample Problem 1: Coloring a Game Board

This problem places a conceptual understanding of the meaning of several fractions into an engaging context for children. Here's the problem:

Guess what? You have been asked to color some game boards for your class!

1. Color in this game board so that 1/2 of the boxes are red and 1/4 are blue.
2. Color the rest of the boxes yellow.

Possible Responses:

A typical response at this stage will be to color half of the boxes red. There are, however, a variety of ways in which a student can respond correctly. Some will color half of *each* box red; this strategy is likely to present difficulty when moving to the next direction, "Color 1/4 of the boxes blue." Some students will divide the whole in half, coloring either the top or bottom row red, or any two adjacent columns red. Again, the next direction will typically prove to be more difficult.

A more sophisticated response will have correctly colored half the boxes red. This student may interpret the next direction as "1/4 of the rest blue" and color in one square blue. If the student has misinterpreted the direction, his conceptual understanding of the fraction may still be sound. Only probing questions can determine this.

The most sophisticated response at this developmental level will correctly interpret all of the directions as fractional parts of the whole array shown. These students should be asked if they can name the fraction that states how much of the board is yellow, and how they know that. Again, be prepared for a variety of ways students can correctly respond to this question, but a correct solution will have the equivalent of four boxes colored red, two boxes colored blue, and two boxes colored yellow.

Extending the Problem

Because the term *half* is commonly used in our language as a quantity that can be any part of a whole, not specifically as one of two equal parts of a whole as it is used in mathematics, it is necessary to follow up with questions designed to get to the heart of the students' understanding of the meaning of the fraction 1/2. Ask them:

1. How many boxes did you color blue?
2. How did you figure out how many boxes to color red?

The problem can be continued by giving students a new game board with new instructions. Students in this extension of the problem are not given a game board that easily lends itself to the given fraction.

1. Color 1/3 of this game board green. You can change the game board if you need to.

Possible Responses to this Extension:

- A typical, but incorrect, response would be to color three boxes green.
- A more sophisticated response can be found in several variations. Some students may ignore the vertical lines, creating two new ones that effectively divide the board into thirds, and color one-third green. Another more sophisticated response would be to color 1/3 of each box on the game board green, thus presenting themselves with many new "wholes" for which 1/3 makes sense to them.
- The most sophisticated response is one in which students add boxes to the game board to give themselves a total number of boxes such as 9 or 12, which they would then use to color the appropriate number green. These students are showing readiness for the concept of common denominators necessary to further operations with fractions. These more sophisticated students, who uncover various methods while working on this problem, can also be asked, "Is there another way to color 1/3 of the game board green?"

(Cont'd.)

Quality Learning Problems: Skills in Context

Finally, prompt all students to circle the game board that was easiest for them and explain why. It is important to ask this question even of students who had difficulty with the first board. In responding, they will reinforce what they do already know about fractions. Some may actually answer this question in the reverse, then state why the other board was difficult. Those students are identifying for themselves what they still need to learn about the meaning of a fraction.

Grades 2/3

Sample Problem 2: Patterns and Their Rules

This problem allows for multiple solution strategies as well as alternative solutions that can still meet the necessary criteria. It provides a window into a child's thinking and approach to a problem. In addition, a link between patterns and data is incorporated, important to later conceptual development in algebraic thinking and functions. The entire problem, with all of its steps, can be found on Worksheet 5.2. It can be administered all at once or broken into parts, similar to the discussion that follows:

Part 1

Someone is beginning a pattern with these shapes:

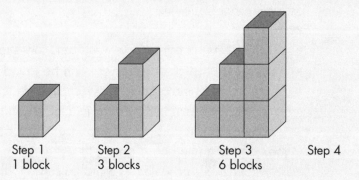

Step 1
1 block

Step 2
3 blocks

Step 3
6 blocks

Step 4

Use your cubes to create the shapes, or think about the shapes that you see.

1. Draw a shape that could be step 4 in the pattern.
2. In words, make up a rule for your pattern so that someone could continue building or drawing shapes that fit the pattern.

Possible Responses:
- A typical response to this problem is one in which the shape is built or drawn, then the total blocks are counted; the rule is not phrased in

general terms, but states numbers ("First one high, then two high," and so on); the location of the next "tower" is not specified.

- A more sophisticated response will be characterized by a consideration of the number of blocks in the next tower before the student builds (if a student chooses to build); this is a rule that generalizes at least to the extent of "one more" and mentions location of the new block or blocks.

There are a variety of ways for a student to arrive at the anticipated solution of "ten blocks" and a general rule that will describe the pattern. Some may consider the shape as a sequence of vertical towers; the tower heights are counting numbers 1, 2, 3, 4. These students start at the beginning for each new step. Others may see an entire shape repeated, with the addition of one tower that is one block higher. Still others may view each shape in a horizontal fashion: "The bottom layer has one block, then two, then three." All of these approaches can yield a viable pattern rule.

It is conceivable that a student will give an unexpected response. In this problem, a student may construct the next vertical tower with only two blocks, for instance. Perhaps the solution looks even more random than that. Questioning and prompting is in order in this situation, as pattern problems often lend themselves to multiple solutions. The correct answer to this problem need not be "ten blocks," although that is the expected and probably most common answer. The correct solution is one in which the same pattern rule "works" from step 1 to step 2, from step 2 to step 3, and so on.

When asked to construct patterns, the best response from a student is one in which the rule explains each step in the sequence that has been given and is stated in general terms so that it describes any step in a sequence of indeterminate length. Students at this level think in concrete terms, and they will need prompting to arrive at a more general description of their pattern.

Part 2

3. Fill in the rest of this chart with the data from your pattern.

Step	Number of blocks in this shape
1	1
2	3
3	6
4	
5	

(Cont'd.)

4. How many patterns can you find in the data chart? Describe them.

Possible Responses:
Some students will see patterns more readily in this format than they might in the actual structures.

- A typical response here will address the numbers in each column, without linking the two columns. "The steps go up by one, and the number of blocks go up by different amounts."
- A more sophisticated response will notice the "pattern within the pattern"; the number of blocks increase by one more each time (first by two, then by three, then by four, then by five), assuming the student created a ten-block shape in step 4.

A student who created an appropriate shape for step 4 using something other than ten blocks often shows more sophistication also when seeing a "pattern within the pattern" in the second column of the data chart.

Part 3

5. Continue filling in numbers in the data chart based on the patterns you saw.

Step	Number of blocks in this shape
1	1
2	3
3	6
4	
5	
6	
7	

Students who have difficulty recognizing the patterns within the data table should be encouraged to return to the more concrete framework of building the structures with manipulatives.

Name _____   Date _____

Patterns and Their Rules

Directions: Use the pictures to help you to answer the questions that follow.

Part 1

Someone is beginning a pattern with these shapes.

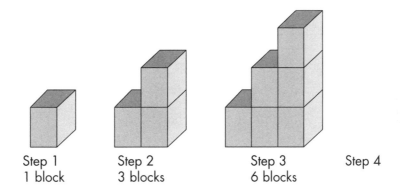

Step 1
1 block

Step 2
3 blocks

Step 3
6 blocks

Step 4

Use your cubes to create the shapes, or think about the shapes that you see.

1. Draw a shape that could be step 4 in the pattern.

2. In words, make up a rule for your pattern so that someone could continue building or drawing shapes that fit the pattern.

Patterns and Their Rules *(Cont'd.)*

Part 2

3. Fill in the rest of this chart with the data from your pattern.

Step	Number of blocks in this shape
1	1
2	3
3	6
4	
5	

4. How many patterns can you find in the data chart? Describe them.

Part 3

5. Continue filling in numbers in the data chart based on the patterns you saw.

Step	Number of blocks in this shape
1	1
2	3
3	6
4	
5	
6	
7	

Grades 2/3

Sample Problem 3:
How Much Money Do I Need?

This problem places computation in a real-world context. The use of "coins" as manipulatives makes the problem more accessible to all students. The initial set of "money" to be given to each student may be in dollar bills, or in coins, depending on the experience students have with money; this problem can be differentiated if different groups are given varied configurations of the initial $2.00. To prepare, a "store" of supplies must be set up, and there must be $2.00 in play money available for each student. Copies of Worksheet 5.3 should be made for each student.

Possible Responses:
- A typical solution to this problem is one in which students actually separate coins, in the exact amounts, into piles for each of the required items. To retain a record of this strategy, the teacher could ask students to paste the coins in groups on another piece of paper, then label the item that it represents. These students will require the precise coins necessary; they will likely be asking for nickels, unless the exact amount of necessary nickels has been provided.
- A more sophisticated solution may take several forms. A single "pile" of money may be used to represent an accumulated total. Some of these students may prefer to perform the calculations necessary without using the manipulatives, or may only use the manipulatives in the areas of difficulty (the stickers).
- The most sophisticated solution is one in which the student can easily complete the problem with two dollar bills provided initially. These students will accumulate a total, even though they may need to "make change" with manipulatives.

The ultimate decision of whether or not to buy some of the extra supplies adds interest to the problem. If a child chooses to get additional supplies, he must show that the money he has left is sufficient. The variety of pricing schemes is intentional. Students will have the most difficulty with determining how much their stickers will cost.

This problem becomes more authentic if a subsequent class activity is conducted in which these supplies are actually used.

Name _____ Date _____

How Much Money Do I Need?

Directions: Read carefully and follow the instructions.

1. Each of you has $2.00 to spend to buy materials for today's lesson. You will need:

   2 markers in different colors

   1 page of cutouts

   6 stickers

   1 piece of poster paper

2. Here is the price list for supplies:

   Markers. . . . . 15 cents each

   Cutouts . . . . 10 cents for each page

   Stickers . . . . 3 for 25 cents

   Poster paper. . . . . 20 cents each

   Glitter . . . . .30 cents each package

   Ribbon .5 cents each piece

You may buy more supplies if you have enough money.

How Much Money
Do I Need? *(Cont'd.)*

3. Answer the following questions:

a. How are you going to spend your $2.00?

b. Will you have money left over?

Sample Problem 1:
Designing a Game Board

This problem requires students to use the concepts of equivalent fractions and common denominators in a context different from addition and subtraction of fractions. Provide plenty of construction paper, rulers, graph paper, circle templates, triangles, and the like for students to consider in their construction of the game board.

Possible Responses:
The correct response to this question is "2/15." Only students who completed a board with 15 (or 30) subdivisions will be able to give this response. Student responses here, however, should be considered in light of their game boards. A student who has labeled 1/4 of his or her game board "Flat Tire" has not met the specifications of the original directions, but would be correct in responding "1/4" to this question. Of significance will be the responses to the second part of the question. Students will be defining the meaning of a fraction in their own words.

- In a typical solution, the first difficulty a student will face in this problem is the selection of a shape for the game board. Students will typically begin with a rectangle or a circle, as they have often seen fractional representations using these constructs. Most will abandon the circle while still dealing with 2/5.
- The next difficulty will be in the determination of 1/3 of the board. Typically, a student may consider 1/3 of the remaining space (after coloring red). Since this simplifies the original problem, this is an incorrect response, and the student should be encouraged to reread the directions for the game board. This student is likely to say, in response to question 3f, that the hardest part of the problem was that it "used fractions."
- A more sophisticated response will employ a structure for which 2/5 is not difficult to determine, and will correctly color that part red. The next direction will be correctly interpreted, but the student will have some difficulty in determining the size of the region needed. This student is likely to overlay divisions of the whole into thirds to find an uncolored region to color green. The student may, at this point, see the need for 15 equal divisions of the board, then begin anew. This student may indicate when responding to question 3f that "starting over" was the greatest

difficulty, still seeing trial and error as the most viable way to approach the problem.

- The most sophisticated response is one in which the student takes all fractions into account at the beginning of the problem, devising a game board that is easily divided into 15 (or 30) regions. When responding to question 3f, this student can articulate the difficulty as being the uncommon denominators.

Worksheet 5.4

Design a Game Board

1. You are a game-board designer for a famous board game company. The game will be called "Red Light, Green Light." You have been asked to design a board so that

   a. 2/5 of it is colored red.

   b. 1/3 of it is colored green.

   c. Of the space remaining, 1/2 is labeled "Flat Tire" and 1/2 is labeled "Pit Stop."

2. Create your game board.

3. After you complete your game board, answer the questions that follow.

   a. What shape is your game board? Why did you pick that shape?

   b. When you needed to color 2/5 of it red, how did you figure out what 2/5 of your board would be?

Name _____   Date _____

Design a Game Board *(Cont'd.)*

c. How do you know for sure that 1/3 of your game board is green?

d. Circle the thing that is the biggest part of your game board.

   Red   Green   Flat Tire   Pit Stop

e. What fractional part of your board is labeled "Flat Tire"? How do you know?

f. What do you think was the hardest part of designing this game board, and why do you think it was so hard?

Sample Problem 2: Allowance Planning

This problem involves adding amounts of money. The lists will be unwieldy enough (fifteen amounts to add) that students will have to devise a strategy to make the process manageable. The problem is interesting enough to fourth or fifth graders that they are likely to be motivated to do this, and the results are surprising. The classic problem from which this arises uses an entire month, but only the most persevering students may be willing to tackle that.

Preparation: Students will need copies of Worksheet 5.5.

The scaffolding in the first question is designed to clarify the three situations presented. A student who has difficulty with this section will be very frustrated by the remainder of this problem. You may want to go over these answers, and the process used to get them, before students continue.

The Problem:

Pretend you have a choice of three plans for your allowance for the first half of November (beginning November 1 and ending November 15).

Choice A.   You get $2.50 every day

Choice B.   Each day you get the same number of dollars as the date (on the 1st you get $1.00, on the 2nd you get $2.00, and so on).

Choice C.   On the 1st you get $.01, and each day you get double what you received the day before.

1. How much would you get on November 4th if you picked:

   Choice A?
   Choice B?
   Choice C?

2. Which plan sounds like it would make you the most money all together?

Possible Responses:

A typical response here is either Choice A or Choice B. In fact, if a student selects Choice C at this stage, it is likely to be due to a knack for recognizing a trick question.

Continuing the Problem:

3. Now, do the math. Which plan will give you the most money all together during the first fifteen days of November?

Possible Responses:
The computations in this section should yield the following results:

> Choice A: $37.50
> Choice B: $120.00
> Choice C: $327.67

It is in the strategies that students use to compute these results that the sophistication of their skills will become apparent.

- A typical strategy is one in which the students list the fifteen numbers to be added; it will be easily recognizable as they work on the totals for Choice A and Choice B. The problem of "lining up the decimal points" will grow as the list gets longer.
- A more sophisticated strategy involves grouping these amounts to get subtotals, before finding the final total. These students, however, are likely to miscount—to add too many or too few numbers—unless they have organized their computations.
- The most sophisticated strategy will be revealed in the work for Choice C. Not only will students have to add fifteen amounts, they will also have to double a previous amount along the way. Organization of the computations, and their sequencing, is important to this result.

It is important to discuss with children how they feel about using a calculator for this problem; although the calculator makes short work of the computation, it will not solve the issues around organizing one's work such that the person solving the problem knows his or her own logical sequence and can follow it.

The Rest of the Problem

4. How much more would it be than if you chose the other two plans?

The solution for this question involves two subtractions and is dependent on the answers found in the previous question. If the answers were correct for Choice A, B, and C, the answers to this problem are that Choice C yields $290.17 more than Choice A and $107.67 more than Choice B. Student responses are correct, however, if they exhibit correct subtraction of incorrect computations from question 3 in this problem.

Worksheet 5.5

Allowance Planning

Directions:

Pretend you have a choice of three plans for your allowance for the first half of November (beginning November 1 and ending November 15).

Choice A.   You get $2.50 every day.

Choice B.   Each day you get the same number of dollars as the date (on the 1st you get $1.00, on the 2nd you get $2.00, and so on).

Choice C.   On the 1st you get $.01, and each day you get double what you received the day before.

1. How much would you get on November 4th if you picked:

   Choice A?

   Choice B?

   Choice C?

2. Which plan sounds like it would make you the most money all together?

Allowance Planning *(Cont'd.)*

3. Now, do the math. Which plan will give you the most money all together during the first 15 days of November?

4. How much more would it be than if you chose the other two plans?

Writing in Math

Although not a "problem" per se, the use of writing prompts in math is another way of tapping or deepening a student's understanding. Switching contexts from numbers to words, and purposes from solving to discussing, can uncover levels of confusion or insight that might not be reached, even by the most open-ended problem or well-constructed activity. To get more from math writing involves pushing beyond "explain how" prompts ("how you determined your answer," "how you solved the problem," "how you decided what steps to take") to prompts that reveal students' dispositions and attitudes at the same time that they evidence content understandings.

Survey questions—familiar in language arts, in which students are asked when they like to read, what books or authors are their favorites, how often they read, what genre of writing they are most comfortable with, and when or why they write by choice—can be used in math as well to help students think about the times in real life when they have needed math that they've learned, or the kind of math problems that they most enjoy, or what world problems might be solved if only the people involved were more aware of math. Interviews with class peers can also provide interesting and important insights.

Asking students to communicate their understanding of how a mathematical model can describe a situation in the real world can quickly tap both content and attitude. For example, in the "problem" that follows—most appropriate for upper elementary students who have some exposure to line graphs—labels and scale are purposefully left off of the graph so that students can relate the graph to many different things from their own life experiences. It is in the story that the students' attitudes and understanding are revealed.

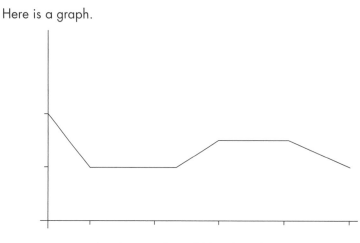

Here is a graph.

Write a story that the graph could be describing.

Creative writing prompts—like describing a world where the primary means of communication was math; or explaining what math is to an alien; or designing a children's book to help primary students understand numbers, counting, or value—are all activities with writing as a medium and math as content. How each of these activities is approached and completed can provide volumes of information about students' dispositions toward, and understanding of, math.

To use writing doesn't necessarily mean that lots of words are necessary. In fact, in some cases, intense writing may actually get in the way of students' being able to communicate clearly, especially if writing is not a strength. Creating math images; using metaphors; comparing math to colors, animals, rooms in a house, songs, or a season of the year—all these can provide a powerful glimpse of student attitudes and dispositions. Verbal responses and class or small-group discussions can also be used where writing may not be the optimum medium for communication.

Responses to these kinds of prompts cannot be categorized as typical or sophisticated, any more than they can be classified as correct or incorrect. It is common, however, for students to reveal frustration and a lack of confidence about the importance of getting the correct answer in mathematics. Teachers may then modify instruction to emphasize that the process is equally important.

Students will also tend to reveal information about their personal learning styles through the explanation of a metaphor. When students are left to choose from a range of objects or ideas to complete an image of math, their responses can very often reveal information related to preferred intelligence or learning style. A student who selects a song, for example, might be drawn to developing conceptual understandings in mathematics through rhythmic or musical representations, whereas a student who selects a sport not only is revealing this as an area of interest for future differentiation but may also learn best through kinesthetic means.

Writing prompts like those described in this section are also useful as part of a pre/post assessment used with students at the beginning of the school year and then revisited at the end of the school year, to measure the change in students' dispositions about and understanding of mathematics.

Some Additional Problems

Me and My Shadow

Students work in pairs, with a shaded lamp and measuring tape. On the wall, student A measures the height of student B and marks it with a piece of masking tape. Student A then measures the height of student B's shadow as that student walks one step, two steps, three steps from the wall (toward the light).

1. What happened as your partner moved closer to the light?
2. What do you think would have happened if your partner had taken smaller steps? Bigger steps?
3. What do you think would happen if you moved farther from the light?
4. What measurement rule seems to be in operation here?

Watching Time Elapse

Students push a stick into the ground or make a stand that will hold the stick upright. They mark the spot on the ground where the stick's shadow is pointing, recording the time. Students record and mark new positions at two other times during the day. They make a prediction about where the shadow will be pointing when school ends, how long would it take for the shadow to return to where it started, and so on—and why they think so.

Meet Me Halfway

Students explore the concept of "halfway" by devising strategies for moving halfway from one place to another and then halfway again. They share their strategies and results, eventually coming up with a definition of "halfway."

Surveys

What would you like to know about what the other students in your class like or dislike?

1. Select something you would like to know about.
2. Create a survey question, with several choices for people to choose from.
3. Conduct your survey and record the data.
4. Decide on the best way to present this data to the class with a data chart and with a graph.
   - Which kind of graph have you chosen for your data? Why?
   - Which shows your survey results better, the data chart or your graph? What is it about the representation that makes you feel it is better?

- Could someone look at your graph and have a different interpretation of the data you collected?
5. After you have seen the presentations of your classmates, answer these questions:
    a. What was different about graphs in which people had more choices when they answered someone's survey question? How might that affect how you interpret the data?
    b. Can a graph of data be misleading?
    c. What kinds of things can be done with a graph or data chart so that it still tells the truth, but makes people think in a different way about the data?
    d. In what kinds of situations would creating a data chart or graph that makes people think in a different way about the data make sense?

Conclusion

Although problems may not be the "answer" for everyone, they can serve to enhance math teaching and learning by providing a reason and context for learning targeted skills and concepts. The ultimate power of problems lies in the conversations that they can inspire and the insight they can provoke—for both students and teachers.

Part Two

Lessons and Activities

Chapter 6

Patterns

What is the difference between a pattern and an event? How can patterns help to solve problems? Where do patterns have an important presence in real life? These are the questions that inform the lessons that follow, connecting patterns as exercises to a focus that makes them meaningful. Without an understanding of patterns and their effects, without an ability to recognize patterns in the world around them, life is little more than a series of independent and disconnected events. For today's students to successfully approach tomorrow's issues, they will need the ability to recognize, analyze, and use patterns.

In this chapter, teacher-directed and student-centered lessons are versions of one another, covering much of the same content, but using different teaching practices, strategies, and approaches. Teacher-directed and student-centered lessons often use the same worksheets, but the practices that contextualize them are different.

Patterns – Grades K/1

Teacher-Directed Lesson

This lesson is teacher-directed, but includes opportunities for students to show what they know in a variety of modalities, from tactile and kinesthetic, to visual, to auditory and linguistic. The teacher guides students through a series of learning opportunities that lead to their understanding of patterns and the rules that establish them.

Lesson Outcome	Students will recognize, articulate, and generate patterns.
NCTM Standards	Algebra Standard: Understand patterns, relations, and functions.
Materials Needed	Pattern blocks or counters, tape recorder, recording of "The Hokey Pokey," student worksheets: 6.1: What's the Pattern? 6.2: Patterns with Partners 6.3: My Very Own Pattern
Time Required	3 hours 20 minutes Can be broken up and implemented over a few days: 45 minutes (1–6); 45 minutes (7–10); 20 minutes (Student Worksheet 6.1); 45 minutes (Student Worksheet 6.2); 45 minutes (Student Worksheet 6.3)
Teaching Practices	Scaffolded questions; multimodal activities; learning opportunities designed to move from concrete to abstract, from lower to higher levels of Bloom's Taxonomy.
Assessment Purposes	Diagnostic – Initial questions about what a pattern is. Formative – Student responses to questions 3, 5, 6, 7b, 8, 11, 12, and 15; Student Worksheet 6.2 Summative – Student Worksheet 6.3

Teaching Guide

1. Ask students what they think "pattern" means and record their answers. Instead of responding to the answers students give, redirect to another student. A conversation might proceed like this:

Teacher: I am going to make a pattern. Who can tell me what that word means?

Joey: A pattern is like a rule.

Teacher: Shantay, what do you think about what Joey just said? Do you have another thing you can say about a pattern?

Shantay: A pattern is like a rule.

Teacher: Joelle, what do you think about what Joey and Shantay just said?

2. Create a pattern by lining students up at the board, according to the rule: boy, girl, boy, girl . . . (but do not reveal the rule to the students).

3. Ask students about what the pattern could be, once again turning responses back to students. When students have agreed on the rule, ask them if the rule could be continued using all the students in the class, ending with the question, "How do you know?"

4. At this point, depending on how well the students have understood, explained, and extended both the pattern and the rule, it may be best to use the children themselves to create another, different pattern. Some possibilities for new patterns are

   Shirt colors

   Two girls, one boy

   Hair or eye color, other features

5. Ask students to describe or explain the pattern rule.

6. Show students a pattern of objects, sequencing by color:

   Red counter, blue counter, yellow counter, and so on

   Again, question students about the rule for the pattern.

7. The Hokey-Pokey:

   a. Play "The Hokey Pokey" and have students move according to the instructions in the song.

   b. Ask students to listen for and tell the patterns in the Hokey-Pokey dance:

   (Right, left . . . in, out, in, shake it all about . . .)

8. Show students a new "counter" pattern:

> 1 red counter, 2 blue counters, 3 red counters, 4 blue counters
>
> Ask: What color comes next? How many should I use?
>
> The following is a pattern within a pattern. Identify both with the students.

Pattern 1:	Red	Blue	Red	Blue	[Red]
Pattern 2:	1	2	3	4	[5]

9. Discuss with students possible ways to state the rule, again turning responses back to the students.

10. Present students with a number pattern.

    a. List numbers from 1 through 10 on the board.

    b. Ask students how they form a pattern.

    c. Circle the numbers that would be the equivalent of skip-counting by two, asking students for various ways to describe that pattern.

11. Distribute *Student Worksheet 6.1: What's the Pattern?*

12. Have students practice prepared patterns at tables.

    a. Students determine what should go next in the pattern and state the rule.

    b. Check for understanding at each table, assessing the progress of students' understanding, and gathering anecdotal evidence. Ask students, "How did you know?"

13. Distribute *Student Worksheet 6.2: Patterns with Partners.*

    a. Students work in pairs or small groups to create a pattern for the class to guess.

    b. Circulate to all of the groups, encouraging students to articulate a pattern rule as they create their pattern.

14. Students share their patterns with the class.

15. Ask students, "What is important to think about when we make a pattern rule?"

    Sample student responses:

    > The pattern stays the same.
    >
    > It makes the pattern.
    >
    > It works.

16. Distribute *Student Worksheet 6.3: My Very Own Pattern.*

    a. Students individually create a pattern for a pattern board that will be displayed in class.

    b. Save this as evidence of learning.

Name _____ Date _____

What's the Pattern?

Finish the patterns.

☺ ☆ ☆ ☺ ☆ ☆ ____

▢ △ ▢ △ ▢ △ ____

A A B A A B A A B ____ ____

1 2 2 2 1 2 2 2 1 2 ____ ____

Names _____   Date _____

_____

_____

Patterns with Partners

Work with one or two partners to create a pattern for the class to guess.

What rule did you use?

_____

_____

_____

_____

_____

Name _____ Date _____

My Very Own Pattern

On your own, create a pattern for the pattern board that will be displayed in class.

Patterns – Grades K/1

Student-Centered Lesson

This lesson uses many of the same activities as the previous, teacher-directed lesson, but it is divided into three parts, with the learning opportunities contained inside differentiated learning centers, allowing students to achieve a greater degree of independence as they develop their understanding of patterns and their rules.

Lesson Outcomes	Students will recognize, articulate, and generate patterns.
	Students will explain their thinking and reasoning about the patterns they experience and create.
NCTM Standards	Algebra Standard:
	Understand patterns, relations, and functions.
	Representation Standard:
	Create and use representations to organize, record, and communicate mathematical ideas.
	Communication Standard
	Communicate their mathematical thinking coherently and clearly to peers, teachers, and others.
	Use the language of mathematics to express mathematical ideas precisely.
Materials Needed	Pattern blocks or counters, tangrams, tape-recorder and headphones, recordings of "The Hokey Pokey," poems with rhyme patterns or of a Dr. Seuss story, preprinted or drawn patterns, student worksheets:
	6.4: Patterns with Color (Center 1)
	6.5: Patterns with Sound (Center 2)
	6.6: Working with Shapes (Center 3)
	6.7: Creating a Joint Pattern (Center 4)
	6.8: My Special Pattern

Time Required	Part 1 – 45 minutes
	Part 2 – 1 hour 30 minutes
	Part 3 – 1 hour
	Part 4 – 20 minutes
Teaching Practices Illustrated	Differentiated learning centers; scaffolded questions; multimodal activities; learning opportunities designed to move from concrete to abstract, from lower to higher levels of Bloom's Taxonomy; incorporation of student reflection
Assessment Purposes	Diagnostic – Part 1
	Formative – Parts 2 and 3
	Summative – Part 4

Teaching Guide

Part 1 – Activating Prior Knowledge

1. Question students as to what they think "pattern" means and record their answers. Use this list as a diagnostic measure of the class's knowledge of patterns. Instead of responding to the answers students give, redirect to another student; for example:

Teacher: I am going to make a pattern. Who can tell me what that word means?

Joey: A pattern is like a rule.

Teacher: Shantay, what do you think about what Joey just said? Do you have another thing you can say about a pattern?

Shantay: A pattern is like a rule.

Teacher: Joelle, what do you think about what Joey and Shantay just said?

2. Create a pattern by lining students up at the board, according to the rule: boy, girl, boy, girl . . . (but do not reveal the rule to the students).

3. Question students about what the pattern could be, once again turning responses back to students to clarify and establish a common

rule. When students have agreed on the rule, ask if the rule could be continued using all the students in the class, ending with the question, "How do you know?"

4. At this point, depending on how well the students understood, explain and extend both the pattern and the rule. It's optional to again use the children themselves, creating a pattern that is different. Some possibilities for new patterns are

Shirt colors

Two girls, one boy

Hair or eye color, other features

Again, students articulate the pattern rule.

Part 2 – Differentiated Learning Centers (Formative Assessment Opportunities)

Inside each learning center, activities are designed to meet the needs of different students' abilities. Students are guided to appropriate activities based on their participation in, and understanding during, Part 1 of this lesson. Providing personal maps that lead students to activities at each center is a strategy that can help avoid students' associating themselves with "high," "average," or "low" activities. This also allows the teacher to customize the experience for each student, based on learning style as well as understanding of the pattern lesson so far.

1. Introduce four centers and explain to students that they will be visiting each center and doing the activities they find there.

Center 1: Visual

Distribute *Student Worksheet 6.4: Patterns with Color.*

Color Sequence Activity I (average) – Pattern of Objects, Sequencing by Color

Sample Pattern: red counter, blue counter, yellow counter, and so on.
   Students:

a. Create a sample pattern.

b. Draw and color in the sample pattern.

c. Articulate the rule in a conference with the teacher.

Color Sequence Activity II (high) – Pattern Within a Pattern, Sequencing by Color

Sample Pattern: One red counter, two blue counters, three red counters, four blue counters.

Students:

a. Continue the pattern.

b. Draw and color in the sample.

c. Articulate the rule in a conference with the teacher.

Color Sequence Activity III (low) – Matching Patterns

Students:

a. Match counters to preprinted or predrawn patterns.

b. Copy and color or attach sample.

c. Articulate the rule in a conference with the teacher.

Center 2: Auditory

Distribute *Student Worksheet 6.5: Patterns with Sound.*

Listen for Patterns I (low) – Recording of Rhythm Patterns (tapping, clapping, and so on)

Students:

a. Listen to a recording of rhythm patterns.

b. Imitate the pattern for the teacher, who documents it on Student Worksheet 6.5 if student is unable to do so.

c. Describe its rule.

d. Share the pattern and rule with the teacher.

Listen for Patterns II (average) – Recording of "The Hokey Pokey"

Students:

a. Listen to a recording of "The Hokey Pokey."

b. Identify patterns in the Hokey -Pokey (right, left . . . in, out, in, shake it all about . . .).

c. Draw, write, or tape-record a pattern when they hear it.

d. Describe the pattern's rule.

e. Share the pattern and rule with the teacher.

Listen for Patterns III (high) – Recording of Poetry or Dr. Seuss Stories
Students:

    a. Listen to recordings of poetry or Dr. Seuss stories.

    b. Identify patterns in the poetry or story.

    c. Draw, write, or tape-record a pattern when they hear it.

    d. Describe the pattern's rule.

    e. Share the pattern and rule with the teacher.

Center 3: Tactile/Kinesthetic
Distribute *Student Worksheet 6.6: Working with Shapes.*

Creating Patterns I (low) – Using Colored Shapes
Students:

    a. Examine patterns of colored shapes provided.

    b. Re-create the patterns using paper cutouts.

    c. Attach the pattern to a piece of paper.

    d. Share the pattern with the teacher.

Creating Patterns II (average) – Using Colored Counters
Students:

    a. Are provided with a "rule" and colored counters.

    b. Apply the rule to create a pattern using the colored counters.

    c. Draw the pattern that they have created.

    d. Share the drawing with the teacher.

Creating Patterns III (high) – Using Tangrams
Students:

    a. Are provided with a "rule" and use it to create a pattern with geometric shapes.

    b. Draw the pattern they have created.

    c. Share the drawing with the teacher.

Center 4: Interpersonal
Distribute *Student Worksheet 6.7: Creating a Joint Pattern.*

Working Together I (low) – Students with Adults
Students:

a. Work in a small group with the teacher or other adult to create a pattern for the class to guess.

b. Share their patterns with the class.

c. Articulate the pattern's rule.

d. Respond to the teacher question, "What is important to think about when we make a pattern rule?"

Working Together II (average) – Students with Students
Students:

a. Work in pairs or small groups to create a simple pattern for the class to guess.

b. Share their pattern with the class.

c. Articulate the pattern's rule.

d. Respond to the teacher question, "What is important to think about when we make a pattern rule?"

Working Together III (high) – Students with Students
Students:

a. Work in pairs or small groups to create a pattern within a pattern for the class to guess.

b. Share their pattern with the class.

c. Articulate the pattern's rule.

d. Respond to the teacher question, "What is important to think about when we make a pattern rule?"

During center time, the assessment of students' progress can be ongoing. When gathering and assessing anecdotal evidence, use a recording sheet like the one that appears at the end of this lesson.

Part 3 – Culminating Assessment
1. Distribute *Student Worksheet 6.8: My Special Pattern*.
2. Students individually create a pattern and explain its rule for a pattern board that will be displayed in the library or a school showcase.

Patterns with Color (Center 1)

Color Pattern

Pattern Rule

_____

_____

_____

_____

_____

Patterns with Sound (Center 2)

Recorded Pattern

Pattern Rule

_____

_____

_____

_____

_____

Name _____   Date _____

Working with Shapes (Center 3)

Shape Pattern

Pattern Rule

_____

_____

_____

_____

_____

Names _____   Date _____

_____

_____

Creating a Joint Pattern (Center 4)

Partner Pattern

What rule did you use?

_____

_____

_____

_____

_____

My Special Pattern

My Pattern

On your own, create a pattern for the pattern board that will be displayed in class.

What is your pattern's rule?

_____

_____

_____

_____

_____

_____

Teacher Resource

Pattern Centers Recording Sheet

Student Name	Visual		Auditory		Tactile/ Kinesthetic	Evidence Collected
	Pattern	Rule	Pattern	Rule	Pattern Uses Rule	

Patterns – Grades 2/3

Teacher-Directed Lesson

This lesson is designed to build on students' prior knowledge and experience by providing them with opportunities to experience more complex patterns, as well as to think about patterns that they notice in real life. They will practice creating patterns using rules they invent and are given, as well as articulating patterns that they notice and the rules that govern them.

Lesson Outcome	Students will recognize and generate patterns that grow and patterns that repeat. They will identify the rules that govern patterns that they recognize as well as use rules to create their own patterns.
NCTM Standards	Algebra Standard:
	Understand patterns, relations, and functions.
	Connections Standard:
	Recognize and apply mathematics in contexts outside of mathematics.
Materials Needed	Overhead, chart paper, or black/whiteboard; student worksheets:
	6.9: What Do I Know About Patterns?
	6.10: Growing and Repeating Patterns
Time Required	3 hours
	Can be broken down as follows:
	1 (30 minutes);
	2–11 (1 hour);
	12 (15 minutes plus homework);
	13–14 (45 minutes);
	15 (30 minutes)
Teaching Practices	Scaffolded questions, guided and independent practice, class discussions, small group reteaching or review (as needed)

Assessment
Purposes

Diagnostic – Student Worksheet 6.9: What Do I Know About Patterns?

Formative – Student responses during class discussions; Student Worksheets 6.10, 6.11.

Summative – Student Worksheet 6.12: What Have I Learned About Patterns?

Teaching Guide

1. Distribute *Student Worksheet 6.9: What Do I Know About Patterns?* Have students complete the worksheet individually.

2. Review student responses for accuracy, understanding, and self-perception. If necessary, engage individuals or small groups in reteaching or review of basic concepts of patterns (see K/1 lesson for possible activities).

3. On an overhead, chart paper, or a black/whiteboard, show students the following pattern:

   2, 2, 4, 6, 8, 10, 12, 14, 16, 18, 20

4. Explain to students that this kind of pattern is a *growing pattern* because the pattern keeps getting bigger, growing every time the pattern repeats.

5. Show students the following pattern:

   2 5 5 2 2 5 5 2 2 5 5

6. Explain that this kind of pattern is called a *repeating pattern* because the pattern repeats in the same order, over and over and over.

7. Show two more patterns:

   1 2 3 4 1 2 3 4 1 2 3 4

8. Ask students which kind of pattern these are. [Answer: repeating]

9. Now, show two more examples (this time, growing patterns):

   27, 28, 29, 30, 31, 32, 33, 34

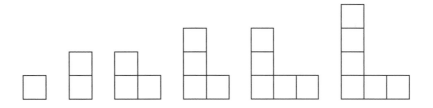

10. Ask students which kinds of patterns these would be.

11. Distribute *Student Worksheet 6.10: Growing and Repeating Patterns.*

12. Patterns in Real Life:

    a. Talk with the class about patterns in real life. Using school as a context, show examples of repeating and growing patterns in the classroom or school. For example, "There is a pattern to what we do during a fire drill: we _____." This is a repeating pattern because we always do the same thing. An example of a growing pattern in school is what happens as students move from grade to grade. Every time you add a year to your age, you move up a grade in school."

    This kind of conversation can bring up opportunities to consider the complexity of real-life patterns and how they can include patterns within patterns and combinations of growing and repeating patterns. For example, questions about classroom routines could illustrate both of these possibilities. If, in order to leave the classroom, students always get into two lines, that is a repeating pattern. If, in order to create the lines, the teacher calls a table or a row at a time to join the line, that is a growing pattern.

    b. Distribute *Student Worksheet 6.11: Patterns Everywhere.*

    c. Have students complete the worksheet for homework. Suggestions: they may look in their neighborhood and notice a pattern in the way the street lights are set, or that there is a stop sign on every corner; they may notice that the homes in their neighborhood all share a similar attribute, or that there are a certain number of homes on a block or on each floor of an apartment building; they may notice a pattern in the way the dinner table is set, or the furniture is arranged; or they may notice a pattern in the way people speak or respond.

13. In small groups, students share their responses, adding to their charts if they so choose, based on what they hear.

14. Create a class chart of "Patterns in Real Life."

    a. Help students determine whether the patterns that they noticed are growing or repeating patterns.

    b. Keep charts of real-life patterns visible and allow students to add to it during the next few weeks or the year.

15. Distribute *Student Worksheet 6.12: What Have I Learned About Patterns?*

    a. Students complete the worksheet individually.

    b. Worksheets can be collected and referenced as evidence of student learning and meta-cognition.

Name _____   Date _____

What Do I Know About Patterns?

Directions: Complete the following patterns and explain the rule that the pattern follows.

A   B   B   A   B   B   A   B   B   ____ ____ ____

                                              Answer: A   B   B

Rule: There is one A and then two Bs, then another A and another two Bs.

Pattern 1.   ▢ △ ▢ △ ▢ △ ____

Pattern Rule: _____

_____

_____

Pattern 2.   11, 12, 13, 14, ___, ___, ___, ___

Pattern Rule: _____

_____

_____

Pattern 3.   1 Red   2 Blue   3 Red   4 Blue   _____

Pattern Rule: _____

_____

_____

What Do I Know About
Patterns? *(Cont'd.)*

Pattern 4.   Create a new pattern that has a different rule. Show the pattern and describe its rule.

```



```

Pattern Rule: _____

_____

_____

1. What is hard for you about working with patterns? What do you think you need help with?

2. What is easy for you about patterns? What do you think you could help others understand about patterns?

Name _____   Date _____

Growing and Repeating Patterns

Complete the following patterns. Say whether it is a growing or repeating pattern and how you know.

	Pattern	Growing or repeating	How do you know?
A.	5, 10, 15, 20, 25, 30, 35, ___, ___, ___		
B.	1, 2, 3, 1, 2, 3, 1, 2, 3, ___, ___, ___		
C.	A B B C C C ___ ___ ___ ___		
D.	$ X $ $ X $ $ $ X ___ ___ ___ ___ ___		
E.	< < < = < < < = < < < = ___ ___ ___ ___		

Create an example of a growing and a repeating pattern:

Growing:

Repeating:

Name _____ Date _____

Patterns Everywhere

Look around you. What patterns do you see? Use this chart to help you keep track of all of them.

Patterns in nature	Patterns in my home	Patterns in my neighborhood	Patterns in people and how they behave

Other patterns that I notice:

Name _____   Date _____

What Have I Learned About Patterns?

Directions:

1. Follow instructions to complete the following patterns.
2. State each pattern's rule.
3. Label each as growing or repeating.

    a. 1, 2, 2, 3, 3, 3, 4, 4, 4, 4,
    Continue this pattern so that it includes the numbers 5 and 6.

    Pattern Rule: _____

    _____

    _____

    Growing or Repeating? _____

    b. 28, 29, 30, 31, ___, 33, 34, 35, 36
    Fill in the missing number.

    Pattern Rule: _____

    _____

    _____

    Growing or Repeating? _____

    c. 5 Yellow   10 Green   15 Yellow   20 Green

    Pattern Rule: _____

    _____

    _____

    Growing or Repeating? _____

Name _____   Date _____

What Have I Learned About
Patterns? *(Cont'd.)*

d. Draw a real-life pattern and describe its rule.

Real-life pattern:

```
┌─────────────────────────────────────────────────────────┐
│                                                         │
│                                                         │
│                                                         │
│                                                         │
│                                                         │
└─────────────────────────────────────────────────────────┘
```

Pattern Rule: _____

_____

_____

Growing or Repeating? _____

4. What is hard for you about recognizing or creating patterns?

5. What do you know about patterns now that you didn't know before we learned about them?

6. What is easy for you about patterns?

7. What are you wondering about patterns now?

Patterns – Grades 2/3

Student-Centered Lesson

This student-centered lesson "covers" the same material as the teacher-centered lesson, but uses constructivist strategies and actively engages students through questioning and investigation. It is designed to build on and deepen students' prior knowledge and experience with patterns, by providing them with opportunities to engage with more complex patterns as well as investigate and appreciate the importance of the patterns that they find in real life.

Lesson Outcome	Students will recognize and generate patterns that grow and patterns that repeat. They will identify the rules that govern patterns they recognize as well as use rules to create their own patterns.
NCTM Standards	Algebra Standard:
	Understand patterns, relations, and functions.
	Reasoning and Proof Standard:
	Make and investigate mathematical conjectures.
	Communication Standard:
	Organize and consolidate their mathematical thinking through communication.
	Communicate their mathematical thinking coherently and clearly to peers, teachers, and others.
	Use the language of mathematics to express mathematical ideas precisely.
	Connections Standard:
	Recognize and apply mathematics in contexts outside of mathematics.
Materials Needed	Overhead or black/whiteboard, chart paper, student worksheets:
	6.13: What Do I Know About Patterns?
	6.14: Growing and Repeating Patterns
	6.15: Noticing Patterns
	6.16: What Have I Learned About Patterns?

Time Required	Part 1: Assessing current understanding (30 minutes);
	Part 2: Deepening skills and understanding (1 hour plus worksheet for homework or class work);
	Part 3: Patterns in real life (45 minutes/break for out-of class assignment/45 minutes);
	Part 4: Assessing learning (30 minutes)
Teaching Practices	Varied questions, scaffolded learning opportunities, constructivist strategies, student inquiry, student reflection, small-group and whole-class work, targeted intervention (as needed)
Assessment Purposes	Diagnostic – Student Worksheet 6.13: What Do I Know About Patterns?
	Formative – Student responses during class discussions and activities, Student Worksheets 6.14 and 6.15
	Summative – Student Worksheet 6.16

Teaching Guide

Part 1: Assessing Current Understanding

1. Distribute *Student Worksheet 6.13: What Do I Know About Patterns?* Have students complete the worksheet individually.

2. Review student responses for accuracy, understanding, and self-perception. If necessary, engage individuals or small groups in reteaching or review of basic concepts of patterns (see K/1 lesson for possible activities).

Part 2: Deepening Skills and Understanding

3. On an overhead, chart paper, or a black/whiteboard, show students the following pattern:

   2, 4, 6, 8, 10, 12, 14, 16, 18, 20

4. Ask students what they notice about the pattern (responses could include skip-counting, counting by twos, numbers getting bigger, every number is two bigger than the one before it).

5. Ask students to explain a rule that would keep the pattern going (answers could include always add two, skip count, or count by two). Write this rule down under the pattern.

6. Show students the following pattern:

   2 5 5 2 2 5 5 2 2 5 5

7. Ask students what they notice about this pattern (responses could include the same two numbers repeat over and over, the numbers go up and then down and then up again, the numbers never change).

8. Have students explain a rule that would keep this pattern going (answers could include repeat (or write) two twos and then two fives over and over; two more twos, then two more fives, then two more twos, and two more fives forever). Write this rule down under the pattern.

9. Ask students "What makes a pattern?" or "What is a pattern?" Use their response to create a definition for the term *pattern*.

10. Post the two patterns and their rules next to one another, underneath the definition of *pattern*. Ask students if they can see any differences between the two patterns. Note their observations.

11. Show students the following pattern:

   a. Ask which of the posted patterns (2, 4, 6, 8, 10, 12, 14, 16, 18, 20 or 2 2 5 5 2 2 5 5 2 2 5 5) this new pattern is most like.

   b. Have students explain what is the same about the pattern they are matching it with (for example, it's most like 2 2 5 5 2 2 5 5 2 2 5 5 because the pattern repeats).

   c. Post the new pattern with the pattern it is most like, along with their comments and observations about the similarities.

12. Show students the following pattern:

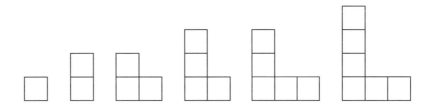

   a. Ask again, which of the two original patterns is most like the new one and why (for example, it belongs with 2, 4, 6, 8, 10, 12 because they both have numbers that get bigger). Post the new pattern with the one it is most like.

   b. Post the new pattern with the pattern it is most like, and also post students' observations and comments.

13. Repeat the process with the following two additional patterns, presenting them one at a time and asking students to put them with the other patterns they are most like and to explain their reasoning.

    27, 28, 29, 30, 31, 32, 33, 34

    1 2 3 4 1 2 3 4 1 2 3 4

    By the end of this discussion, the groups should be:

    A   **B**

    2 2 5 5 2 2 5 5 2 2 5 5   2, 4, 6, 8, 10, 12, 14, 16, 18, 20

    ☺ ☆ ☆ ☺ ☆ ☆ ☺ ☆ ☆

    1 2 3 4 1 2 3 4 1 2 3 4   27, 28, 29, 30, 31, 32, 33, 34

14. Once all patterns are grouped and the reasons documented, ask students to look carefully at what they have said about the reasons the patterns belong together. Based on what they have said, have them articulate a rule for each group. (This can be done as a whole class, or students can break into smaller groups of three or four, articulate the two rules, and write them on chart paper. The rules can be posted and similarities highlighted. A class draft can be formulated by combining what is common.)

15. Show students the terms *growing pattern* and *repeating pattern* and ask them, based on the rules that they just described, which would be a good title for each pattern group. (A = repeating, B = growing.)

16. Distribute *Student Worksheet 6.14: Growing and Repeating Patterns*. This can be assigned as homework or class work. In either case, it should be reviewed for understanding and will provide an opportunity to recognize which students might be having difficulty. These students can then receive one-on-one or small-group targeted instruction.

Part 3: Patterns in Real Life

17. As a class, begin to consider important patterns in real life, beginning with school as a context.

    a. Ask questions that will help students recognize repeating and growing patterns in classroom or school routines or expectations. For example: "What are the things that we always do or say when there is a fire drill?" and when answers are given, "Would you call that a repeating pattern, or a growing pattern?" or "What kind of pattern can you see in the way a student moves from grade to grade?" or "What is the pattern in our writing workshop?" or "What pattern rule helps us leave the classroom in an orderly way?"

b. Use the definitions for *pattern, growing pattern,* and *repeating pattern,* drafted early in the lesson, to help assess whether what is suggested is indeed a pattern—and what kind of pattern it might be.

This kind of conversation can bring up opportunities to consider the complexity of real-life patterns and how they can include patterns within patterns and combinations of growing and repeating patterns. A response to the last question about classroom routines could illustrate both of these possibilities. If, in order to leave the classroom, students always get into two lines, that is a repeating pattern. If, in order to create the lines, the teacher calls a table or a row at a time to join the line, that is a growing pattern.

c. Move to other aspects of students' own lives. What patterns do they recognize when they get ready for school? In other classes or times during the school day? When they return home in the afternoon? Do they experience different patterns on weekends? Continue to use class definitions as criteria to maintain the focus on patterns, refining them if necessary.

18. Discuss categories that patterns might fall into, prompting students to generate at least some of them. Many of the previous questions will lead students to recognize patterns in behavior, but there are many more. Other categories might be patterns in places (this could refer to something as small as the layout of the room or the architecture of buildings to patterns that can be seen in my neighborhood or even city planning), patterns in nature (as small as a cell or patterns in the fur of a pet, the leaves on a tree or make up of a flower, to patterns in individual species or the plant or animal kingdom), patterns in different kinds of writing, patterns in clothing (material, style, seasonal attributes), patterns in music, and the like.

19. From the list that has been generated, have students select a category that they would like to find examples for (it is not necessary that all categories be chosen). Distribute *Student Worksheet 6.15: Noticing Patterns.* Model the use of the worksheet by fitting in some of the examples already discussed. You may want to post the model so that students can use it as a reminder.

20. Allow students two or three days to investigate the category they have selected.

21. When students return with their examples, organize them into groups based on the category they selected and investigated. (If a group is too large, it can be divided into more than one group.) In their category

groups, students will share the patterns that they have discovered. As a group, they will use the class pattern definitions to help them identify strong examples of repeating and growing patterns from those that are shared (the number per group can be limited or the only requirement can be that examples fit the definitions).

22. Have category groups create posters of their examples that can then be displayed.

If these patterns or pattern categories support content or processes in other curriculum areas, the postings can remain and be used or built on throughout the year as an opportunity to establish interdisciplinary connections.

Part 4: Assessing Learning

23. Distribute *Student Worksheet 6.16: What Have I Learned About Patterns?* Have students complete the worksheet individually

Name _____   Date _____

What Do I Know About Patterns?

Directions: Complete the following patterns and explain the rule that the pattern follows.
Example:

A   B   B   A   B   B   A   B   B   ____ ____ ____

Answer: A   B   B

Rule: There is one A and then two Bs, then another A and another two Bs.

Pattern 1.   ☐ △ ☐ △ ☐ △ ____

Pattern Rule: _____

_____

_____

Pattern 2.   11, 12, 13, 14, ___, ___, ___, ___

Pattern Rule: _____

_____

_____

Pattern 3.   1 Red   2 Blue   3 Red   4 Blue   _____

Pattern Rule: _____

_____

_____

What Do I Know About
Patterns? *(Cont'd.)*

Pattern 4. Create a new pattern that has a different rule. Show the pattern in the following space and describe its rule.

```




```

Pattern Rule: _____

_____

_____

1. What is hard for you about working with patterns? What do you think you need help with?

2. What is easy for you about patterns? What do you think you could help others understand about patterns?

Name _____ Date _____

Growing and Repeating Patterns

Complete the following patterns. Say whether it is a growing or repeating pattern and how you know.

	Pattern	Growing or repeating	How do you know?
A.	5, 10, 15, 20, 25, 30, 35, ___, ___, ___		
B.	1, 2, 3, 1, 2, 3, 1, 2, 3, ___, ___, ___		
C.	A B B C C C ___ ___ ___ ___		
D.	$ X $ $ X $ $ $ X ___ ___ ___ ___ ___		
E.	< < < = < < < = < < < = ___ ___ ___ ___		

Create an example of a growing and a repeating pattern:

Growing:

Repeating:

Name _____   Date _____

Noticing Patterns

Think about the category that you have chosen and find patterns to share with the class. When you notice a pattern, write it in your chart. Do you think it might be a growing pattern or a repeating pattern?

Pattern Category:		
Pattern example	**Is it a growing or repeating pattern?**	**What makes me think that?**
	Growing Repeating	
	Growing Repeating	
	Growing Repeating	
	Growing Repeating	

Name _____   Date _____

What Have I Learned About Patterns?

Directions:

1. Follow instructions to complete the following patterns.
2. State each pattern's rule.
3. Label each as growing or repeating.

Pattern 1.   1, 2, 2, 3, 3, 3, 4, 4, 4, 4,
Continue this pattern so that it includes the numbers 5 and 6.

Pattern Rule: _____

_____

_____

Growing or Repeating? _____

Pattern 2.   28, 29, 30, 31, ___, 33, 34, 35, 36

Fill in the missing number.

Pattern Rule: _____

_____

_____

Growing or Repeating? _____

Pattern 3.   5 Yellow   10 Green   15 Yellow   20 Green

Pattern Rule: _____

_____

_____

Growing or Repeating? _____

Name _____ Date _____

What Have I Learned About Patterns? *(Cont'd.)*

Pattern 4.   Draw a real-life pattern and describe its rule.

Real-life pattern:

```

```

Pattern Rule: _____

_____

_____

Growing or Repeating? _____

1. What is hard for you about recognizing or creating patterns?

2. What do you know about patterns now that you didn't know before we learned about them?

3. What is easy for you about patterns?

4. What are you wondering about patterns now?

Patterns – Grades 4/5

Teacher-Directed Lesson

Through completion of a variety of activities, students explore and deepen their understanding of patterns, ultimately making connections to the importance of patterns in problem solving.

Lesson Outcome	Students will recognize and generate patterns that grow and patterns that repeat as well as the rules that govern them, and they will be able to use patterns to help them approach and solve problems.
NCTM Standards	Algebra Standard:
	Understand patterns, relations, and functions.
	Connections Standard
	Recognize and apply mathematics in contexts outside of mathematics.
Materials Needed	Overhead, chart paper, or black/whiteboard; student worksheets:
	6.17: Patterns in Many Places
	6.18: Patterns in Problems
	6.19: The Disappearing Dog Dilemma
Time Required	3.5 hours total – can be broken into smaller timeframes based on individual activities.
Teaching Practices	Scaffolded questions, guided and independent practice, class discussions, small group reteaching or review (as needed)
Assessment Purposes	Formative – Student responses during class discussions; Student Worksheet 6.17
	Summative – Student Worksheets 6.18 and 6.19

Teaching Guide

1. On overhead, chart paper, or black/whiteboard, show a tic-tac-toe grid. Play three different games with three different students while other students watch for patterns in both the grid itself (such as

three X's or O's in a row or column or on a diagonal) or in the strategies for playing the game (for example, if the first person to go starts in the center). After each game, articulate any patterns that you see (or elicit them from the students). Keep a list of the patterns. Ask, "Is this game based on a growing or repeating pattern?" (Repeating = three X's or O's in a row.)

2. Distribute *Worksheet 6.17: Patterns in Many Places.*

   a. In Part 1, Making Boxes, have students divide into pairs to play the connect-the-dots game.

   b. Playing two games, student pairs make note of the game's patterns and pattern rules (take turns, connect dots, last line gets the box, and so on) and identify the kind of pattern on which the game is based (in this case, the game is based on a growing pattern because the lines build boxes and the player with the most boxes wins).

   c. Students share answers as a class.

3. Tell students that you are going to read something aloud and they should listen carefully for the rhythm or beat in the words. (Nursery rhymes like "Little Bo Peep," "Jack and Jill," "Humpty Dumpty," or "Baa Baa Black Sheep," as well as Dr. Seuss stories and *Brown Bear, Brown Bear, What Do You See?* by Bill Martin Jr., work well for this activity.)

   a. Read "Jack and Jill" aloud and show students how to tap or nod the rhythm. Repeat with the same example, this time letting students tap or nod. Model on overhead or chart paper how to document the rhythm, using an X for accented syllables and a dash for unaccented syllables.

   b. Point students to the section of Student Worksheet 6.17 labeled Part 2, Patterns in Rhythm; have students read "Baa Baa Black Sheep" to examine for rhythm. Each student documents the pattern following the teacher's model.

   c. Read "Humpty Dumpty" to students, this time listening for rhyme. Model for students how to record a rhyme scheme using letters. For example, "Humpty Dumpty sat on a wall/Humpty Dumpty had a great fall/All the king's horses and all the king's men/ Couldn't put Humpty together again" has a rhyme pattern of wall/ fall, men/again; this pattern could be represented with the letters a, a, b, b and the rule could be explained as "the last words of the first two sentences rhyme with each other, the last words of the next two sentences rhyme with each other."

   d. Point students to the section of Worksheet 6.17 labeled Part 3, Patterns in Rhyme, and have them read a page from Dr. Seuss's *The Cat in the Hat* or *Green Eggs and Ham* to examine for rhyme pattern. Each student documents the pattern following the teacher's model.

4. Show samples of mandalas, mosaics, or quilts, explaining where the patterns are in each (conduct an online search for images or see the Resources section for Web sites to visit).

   a. Point students to the section of Worksheet 6.17 labeled Part 4, Patterns in Art.

   b. Students examine another example of a mandala, mosaic, or quilt to identify and describe embedded patterns.

   c. On their worksheets, they draw and describe the patterns that they see, including shapes and colors.

5. Collect Worksheet 6.17 and assess for understanding, determining what, if any, additional lessons students might require.

6. Show students the following logic problem:

   Tina, Mark, Jose, George, and Nina are waiting in the lunch line. Can you tell what order the students are standing in?

   • Tina is wearing green sneakers.

   • The teacher always alternates boys and girls in a line.

   • There is no one in line behind Mark.

   • Jose is standing between two girls.

   • George is standing next to someone wearing blue sneakers.

     (Answer: George, Nina, Jose, Tina, Mark)

   a. What pattern was important to solving this logic problem? Was it a growing pattern or a repeating pattern? (Alternating boys and girls = repeating.)

   b. Distribute *Student Worksheet 6.18: Patterns in Problems*.

   c. Students solve the word problems on the worksheet, identifying the patterns that they found and used in the process.

7. Explain that patterns help people solve real-life problems. Discuss with students how detectives look for patterns in evidence to solve a crime, doctors look for patterns in symptoms when they diagnose illness, city planners and architects use patterns in their designs of cities and buildings, advertisers look for patterns in consumer spending, and so on. If desired, ask students to expand this list.

8. Distribute *Student Worksheet 6.19: The Disappearing Dog Dilemma*. Remind students that patterns repeat. Students use their ability to identify patterns to help Detective Harry solve the case.

9. Assess student responses and document evidence of student learning.

Patterns in Many Places

Part 1 – Making Boxes

The purpose of this game is to makes boxes by connecting the dots. Taking turns with your partner, draw a line that connects two dots (diagonal lines are not allowed). If you draw the last line to close off a box, then you get that box and you can put your initial inside it. The person who has the most boxes at the end wins.

Play two games, using your partner's and your own game board. Keep track of the patterns that you see and the patterns that you use as you play.

My notes on the patterns:

What pattern is this game based on?

Is it a growing or repeating pattern?

Patterns in Many Places *(Cont'd.)*

Part 2 – Patterns In Rhythm

Read the following nursery rhyme. Write the rhythm pattern that you hear, using an X for accented syllables and a dash for unaccented syllables. Place the symbols directly over the syllables in the poem.

Baa baa black sheep, have you any wool?

Yes sir, yes sir, three bags full.

One for the master,

One for the dame,

And one for the little boy who lives down the lane.

Part 3 – Patterns In Rhyme

Read a page from *The Cat in the Hat* or *Green Eggs and Ham*.

Write letters to show the rhyme pattern. Be sure to cite the story and the page that you analyzed.

Story/page:

Rhyme pattern:

Part 4 – Patterns In Art

Look at one of the samples. Sketch and describe a pattern that you see.

Sketch

Description:

Name _____   Date _____

Patterns in Problems

Find the pattern and solve the problem.

1. Sara is older than John. John is the baby of the family. Martina, the oldest, is twelve. Jason, the next to youngest, is six. There are three years between each of the children. How old is Sara?

   Sara's age:

   Pattern that helped you:

2. Something is wrong with the wall. It won't fall down, but it looks wrong. Use what you know about patterns to explain to the builder what is wrong with the wall, and how it can be fixed.

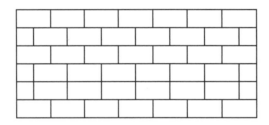

   a. What is wrong with the way the wall looks?

   b. What is the pattern that helps you to know?

   c. How can this be fixed?

Patterns in Problems *(Cont'd.)*

3. As part of their research for a new marketing campaign, an ice cream company surveyed sixty consumers about their flavor preferences.

   - One out of every four people said that they preferred chocolate.
   - One of every three said that their favorite flavor was vanilla.
   - The rest were undecided.

   a. What is the pattern of those preferring chocolate?

   b. If you use this pattern, how many of the sixty people actually liked chocolate best?

   c. How many people preferred vanilla? What pattern did you use?

   d. How many were undecided?

The Disappearing Dog Dilemma

Patternicus, the Smiths' three-year-old performing poodle, is missing. He disappeared a week-and-a-half ago from the kennel where he was being cared for while the Smith family was away on a ski vacation.

The police department has sent their very best detective, Detective Harry, to solve the case. There has never been a case that Harry could not solve. This one, however, is giving him some trouble. There are so many clues that he is having trouble keeping track of them and what they might mean. He needs help looking at the patterns and events to figure out who is really responsible for the dog's disappearance.

Here are his notes:

The dog

1. He is three years old.
2. Every day he eats chicken and chicken treats.
3. For three years, Patternicus performed amazing tricks around the state, earning lots of money and making Mr. and Mrs. Smith famous.
4. Lately, he has refused to perform, leaving audiences angry and costing the Smiths a lot of money in cancellation fees.
5. He behaves badly at home, too: chews furniture and barks.
6. He behaves and performs only for Grandmother Smith, who keeps a supply of his favorite chicken treats in her pocket.

Smith family

1. Two parents, two children (ages seven and nine).
2. The Smith family goes on family ski weekends two times per month, every winter.
3. For the past three years, Grandmother Smith has taken care of Patternicus when the Smith family take their weekend ski trips and any other time the dog was going to be left alone for more than a day.
4. The more she has taken care of the dog, the more Grandmother Smith likes Patternicus.
5. Grandmother Smith refers to Patternicus as "my Patty."
6. Mr. and Mrs. Smith are becoming more and more frustrated with Patternicus's behavior.
7. The Smiths are considering selling Patternicus and have contacted someone who is interested.
8. Grandmother Smith has argued daily with Mr. and Mrs. Smith over the fact that Patternicus may be sold.
9. There have been daily deliveries of chicken to Grandmother Smith's house for the past week.

The Disappearing Dog
Dilemma *(Cont'd.)*

The kennel

1. The only footprints outside the kennel were a single set of dog prints, leading away from the building.
2. Kennel workers said that Patternicus was acting strangely before the disappearance, staring at an open window and barking.

Other

1. When dog prints were run through the crime lab computer, they came back as a perfect match to Patternicus's records.
2. An empty bag of chicken treats was found on the curb outside the kennel.
3. A reward has been offered, but so far no one has responded.

Your analysis:

Who took Patternicus?
What patterns in Detective Harry's notes lead you to believe this?

Patterns – Grades 4/5

Student-Centered Lesson

This student-centered lesson "covers" the same material as the teacher-directed lesson, but deepens the experiences by actively engaging students through questioning and investigation. Interactive learning centers provide the structure for this lesson, designed to build on students' prior knowledge and experience with patterns by providing them with opportunities to engage with more complex patterns as well as investigate and appreciate the importance of patterns in solving real-life problems.

Lesson Outcome	Students will recognize and generate patterns that grow and patterns that repeat. They will identify the rules that govern patterns they recognize as well as use rules to create their own patterns.
NCTM Standards	Algebra Standard:
	Understand patterns, relations, and functions.
	Reasoning and Proof Standard:
	Make and investigate mathematical conjectures.
	Communication Standard:
	Organize and consolidate their mathematical thinking through communication.
	Communicate their mathematical thinking coherently and clearly to peers, teachers, and others.
	Use the language of mathematics to express mathematical ideas precisely.
	Connections Standard:
	Recognize and apply mathematics in contexts outside of mathematics.

Materials Needed	Overhead or black/whiteboard, chart paper, student worksheets:
	6.20: What Do I Already Know About Patterns? (Center 1)
	6.21: Patterns in Games (Center 2)
	6.22: Patterns in Rhythm and Rhyme (Center 3)
	6.23: Patterns in Art (Center 4)
	6.24: Patterns in Problems
	6.25: The Disappearing Dog Dilemma
	6.26: Reflecting on Patterns
Time Required	20 to 60 minutes per center and major activity, depending on activities and experience of students
Teaching Practices	Varied questions, centers, modeling, think-aloud, student inquiry, student reflection, small group and whole class work, targeted intervention (as needed)
Assessment Purposes	Diagnostic – Student Worksheet 6.20
	Formative – Student responses during class discussions and activities, student worksheets for all centers
	Summative – Student Worksheets 6.24, 6.25, and 6.26

Teaching Guide

Introducing Learning Centers

These interactive centers focus on patterns found in math and other contexts. Students explore prior knowledge about identifying and creating patterns and deepen their understanding and skills through completion of a variety of activities, followed by answering related prompts. Centers can be actual physical spaces where students go to complete the activities, or they can be sets of activity materials that students use back in their own workspace. Students cycle through each center until all have been experienced; they should complete Center 1 first and Center 4 last, but other than that, the order is not important. Students should be free to enter and exit a fifth center, the conferencing center, where they can consult with the teacher as needed.

This section describes the four centers and their activities.

Center 1 – What Do I Already Know About Patterns?

Distribute *Student Worksheet 6.20: What Do I Already Know About Patterns?*

a. Students complete the patterns provided in Worksheet 6.10.

b. Assess these for understanding. Assign students to the conference center for review or reteaching as needed.

Center 2 – Patterns in Games

Distribute *Student Worksheet 6.21: Patterns in Games.*

Tic-Tac-Toe

a. In pairs, students play four games of tic-tac-toe as they try to uncover one or more patterns that relate to strategies for playing the game (for example, if the first person to go starts in the center).

b. After each game, students document these patterns on their worksheets to be shared later.

c. Patterns and rules are posted in a chart at the center or in a labeled space in the classroom. If a pattern is already posted, students should not post it again, but should put a tally mark next to the one that is already there.

Making Boxes

a. Students work in pairs to connect dots and make and claim boxes.

b. Playing two games, students make note of the game's patterns and pattern rules (take turns, connect dots, last line gets the box, and so on) and identify the kind of pattern on which the game is based (in this case, the game is based on a growing pattern because the lines build boxes and the player with the most boxes wins).

c. After each game, students document these patterns on their worksheets.

d. Patterns and rules are posted in a chart at the center or in a labeled space in the classroom. If a pattern is already posted, students should not post it again, but should put a tally mark next to the one that is already there.

Center 3 – Patterns in Rhythm and Rhyme

Distribute *Worksheet 6.22: Patterns in Rhythm and Rhyme.* Provide copies of poetry having a strong sense of rhythm as well as strong rhyme schemes. (Nursery rhymes like "Little Bo Peep," "Jack and Jill," "Humpty Dumpty" or "Baa Baa Black Sheep," as well as Dr. Seuss stories and *Brown Bear, Brown Bear, What Do You See?* by Bill Martin Jr., work well for this activity.)

Patterns in Rhythm

a. Post a model of rhythm analysis at the center ("Jack and Jill" is easy to follow; mark an X over all accented syllables and a dash over all unaccented syllables).

b. Working in pairs, students select two examples to examine for rhythm. One partner reads the sample, repeating it over and over without stopping. As the rhythm becomes more apparent, the other partner claps or taps the rhythm that they hear, giving more emphasis to accented words and syllables.

c. Once the rhythm is established, each student documents the pattern according to the model.

Patterns in Rhyme

a. Post a model of rhyme analysis at the center. For example, "Humpty Dumpty sat on a wall/Humpty Dumpty had a great fall/ All the king's horses and all the king's men/Couldn't put Humpty together again" has a rhyme pattern of wall/fall, men/again; this pattern could be represented with the letters a, a, b, b and the rule could be explained as "the last words of the first two sentences rhyme with each other, the last words of the next two sentences rhyme with each other."

b. Working individually, students select two different samples from the center and identify the pattern in the rhyme scheme. They share these with a classmate or meet with the teacher at the conferencing center for feedback.

c. Students analyze "Simple Simon" and work with a partner to respond to questions.

Center 4 – Patterns in Art

1. Distribute *Student Worksheet 6.23: Patterns in Art.* Provide samples of mandalas, mosaics, or quilts for students to examine and work with (conduct an online search for images or see the Resources section for Web sites to visit).

a. Students examine samples of mandalas, mosaics, and quilts to identify and describe embedded patterns. They document their findings on Worksheet 6.23, including shapes and colors.

b. Students select either a preprinted picture or a piece of graph paper and create their own mandala, mosaic, or quilt pattern that incorporates patterns of shape and color. They describe the key patterns in their project and explain the pattern rules that guide them.

c. These designs are posted at the center or in a labeled area of the classroom.

2. How do patterns help in problem solving?

a. After completing the previous centers, in small groups of three or four, students brainstorm real-life situations in which people use patterns to help create, organize, or interpret information. One or two of these can be preposted on a piece of chart paper, titled "Using Patterns in Real Life" (for example, detectives look for patterns in evidence to solve a crime; doctors look for patterns in symptoms when they try to figure out what is wrong with patients; city planners and architects use patterns in their designs of cities and buildings; advertisers look for patterns in consumer spending).

b. As soon as a group has one or two ideas, they can add them to those already posted.

3. Explain to students that the ability to recognize patterns is a powerful problem-solving strategy. Referring to the list that they just created, prompt students to identify specific examples of problems or to extend their list to include problems or problem-solving situations in which recognizing and using patterns would be necessary (or very helpful) to find a solution (for example, solving a crime, curing a disease or stopping an epidemic, winning an election).

4. Use the following simple logic problem as a model for using patterns in problem solving.

Tina, Mark, Jose, George and Nina are waiting in the lunch line. Can you tell what order the students are standing in?

- Tina is wearing green sneakers.
- The teacher always alternates boys and girls in a line.
- There is no one in line behind Mark.
- Jose is standing between two girls.
- George is standing next to someone wearing blue sneakers.

(Answer: George, Nina, Jose, Tina, Mark)

Think aloud to share the process, modeling not only the actual problem solving but the documentation of thinking and pattern identification that students will be encouraged to follow. Ask questions about what pattern was important to solving this logic problem and if it was a growing pattern or a repeating pattern. (Alternating boys and girls = repeating.)

5. Distribute *Student Worksheet 6.24: Patterns in Problems.*

    a. Have students complete this worksheet as class work or assign as homework.

    b. Collect and assess as evidence of learning and to determine if there is a need for review or reteaching, prior to assigning Worksheet 6.25.

6. Distribute *Student Worksheet 6.25: The Disappearing Dog Dilemma.*

    a. Ask students to think about events (a ball game, an argument, a vacation, a birthday party). What do they have in common? (For example: they happen once or once in a while, they are not always predictable, they have an end).

    b. Ask students to think about what they know about patterns. What do they have in common? (For example, they repeat in a regular way, they're predictable, they continue.)

    c. Have students write these definitions at the top of Worksheet 6.25.

7. Have students complete *Worksheet 6.26: Reflecting on Patterns.*

8. Collect and assess Worksheets 6.25 and 6.26 as evidence of student learning and thinking.

Name _____   Date _____

What Do I Already Know About Patterns? (Center 1)

Directions:

1. Follow instructions to complete the following patterns.
2. State each pattern's rule.
3. Label each as growing or repeating.

Pattern 1.   1, 2, 2, 3, 3, 3, 4, 4, 4, 4,
Continue this pattern so that it includes the numbers 5 and 6.

Pattern Rule: _____

_____

_____

Growing or Repeating? _____

Pattern 2.   32, 64, 128, 356, 712, _____, 2848
Fill in the missing number.

Pattern Rule: _____

_____

_____

Growing or Repeating? _____

Pattern 3.   25 Yellow   10 Green   30 Yellow   20 Green

Pattern Rule: _____

_____

_____

Growing or Repeating? _____

What Do I Already Know About Patterns? (Center 1) *(Cont'd.)*

4.   Draw a real-life pattern and describe its rule.

Real-life pattern:

_____

Pattern Rule: _____

_____

_____

Growing or Repeating? _____

5. What is hard for you about recognizing or creating patterns?

6. What is easy for you about patterns?

Patterns in Games (Center 2)

Tic-Tac-Toe

Play four games of tic-tac-toe with a partner. Each time you play, think about the patterns that you see in the game, especially patterns that help you win or stop the other player from winning. Use the chart to help you keep track of your thinking.

Game	Patterns Noticed	Pattern
1		
2		
3		
4		

Post your patterns and rules. If a pattern or rule is already posted, then put a tally mark next to it.

Name _____   Date _____

Patterns in Games (Center 2) *(Cont'd.)*

Making Boxes

The purpose of this game is to makes boxes by connecting the dots. Taking turns with your partner, draw a vertical or horizontal line that connects two dots (diagonal lines are not allowed). If you draw the last line to close off a box, then you get that box and you can put your initial inside it. The person who has the most boxes at the end wins.

Play two games, using your partner's and your own game board. Keep track of the patterns that you see and the patterns that you use as you play.

a. My notes on the patterns:

b. What pattern is this game based on?

c. Is it a growing or repeating pattern?

Post your patterns and rules. If a pattern or rule is already posted, then put a tally mark next to it.

Worksheet 6.22

Patterns in Rhythm and Rhyme (Center 3)

1. Patterns in Rhythm:

   Choose a nursery rhyme from the center. Copy it here and write the rhythm pattern that you hear, using an X for accented syllables and a dash for unaccented syllables. Place the symbols directly over the syllables in the poem.

2. Patterns in Rhyme:

   Choose a page from *The Cat in the Hat* or *Green Eggs and Ham*, or a different nursery rhyme than the one that you just used.

   Follow the model and use letters to show the rhyme pattern here. Be sure to cite the story and the page or the nursery rhyme that you analyzed.

   Story and page:

   Rhyme pattern:

3. Patterns in Rhythm and Rhyme:
   a. Read the entire poem that follows.
   b. Select two stanzas to analyze for rhythm pattern and two different stanzas to analyze for rhyme pattern.

Patterns in Rhythm and Rhyme
(Center 3) *(Cont'd.)*

Simple Simon

Simple Simon met a pieman,
Going to the fair.
Said Simple Simon to the pieman,
"Let me taste your ware."

Said the pieman unto Simon,
"Show me first your penny."
Said Simple Simon to the pieman,
"Indeed I have not any."

Simple Simon went a-fishing,
For to catch a whale;
But all the water he had got
Was in his mother's pail.

Simple Simon went to look,
If plums grew on a thistle;
He pricked his fingers very much,
Which made poor Simon whistle.

He went for water in a sieve,
But soon it all fell through;
And now poor Simple Simon
Bids you all adieu.

4. Talk to a partner about the following two questions:
   a. How does the writer use the pattern to keep the reader's attention?

   b. What does the writer do to the pattern at the end? What effect does that have?

Name _____   Date _____

Patterns in Art (Center 4)

1. Look at one of the samples. Sketch and describe a pattern that you see.

Sketch

Description (Include colors and shapes):

2. Select either a preprinted picture or a piece of graph paper to design your own. Use patterns of shapes and colors to create a mini mosaic, mandala, or quilt design.
   Post your design when you are finished.

Name _____ Date _____

Patterns in Problems

Find the pattern and solve the problem.

1. Sara is older than John. John is the baby of the family. Martina, the oldest, is 12. Jason, the next to youngest, is 6. There are 3 years between each of the children. How old is Sara?

Sara's age:

Pattern that helped you:

2. Something is wrong with the wall. It won't fall down, but it looks wrong. Use what you know about patterns to explain to the builder what is wrong with the wall, and how it can be fixed.

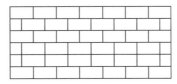

a. What is wrong with the way the wall looks?

b. What is the pattern that helps you to know?

c. How can this be fixed?

Patterns in Problems *(Cont'd.)*

3. As part of their research for a new marketing campaign, an ice cream company surveyed sixty consumers about their flavor preferences.

   • One out of every four people said that they preferred chocolate.
   • One of every three said that their favorite flavor was vanilla.
   • The rest were undecided.

   a. What is the pattern of those preferring chocolate?

   b. If you use this pattern, how many of the sixty people actually liked chocolate best?

   c. How many people preferred vanilla? What pattern did you use?

   d. How many were undecided?

Name _____   Date _____

The Disappearing Dog Dilemma (student-centered version)

Patternicus, the Smiths' three-year-old performing poodle, is missing. He disappeared a week-and-a-half ago from the kennel where he was being cared for while the family was away on a ski vacation.

The police department has sent their very best detective, Detective Harry, to solve the case. There has never been a case that Harry could not solve. This one, however, is giving him some trouble. There are so many clues that he is having trouble keeping track of them and what they might mean. He needs help looking at the patterns and events to figure out who is really responsible for the dog's disappearance

Here are his notes:

The dog

1. He is three years old.
2. Every day he eats chicken and chicken treats.
3. For three years, Patternicus performed amazing tricks around the state, earning lots of money and making Mr. and Mrs. Smith famous.
4. Lately, he has refused to perform, leaving audiences angry and costing the Smiths a lot of money in cancellation fees.
5. He behaves badly at home, too: chews furniture and barks.
6. He behaves and performs only for Grandmother Smith, who keeps a supply of his favorite chicken treats in her pocket.

Smith family

1. Two parents, two children (ages seven and nine).
2. The Smith family goes on family ski weekends two times per month, every winter.
3. For the past three years, Grandmother Smith has taken care of Patternicus when the Smith family takes their weekend ski trips and any other time the dog is going to be left alone for more than a day.
4. The more she has taken care of the dog, the more Grandmother Smith likes Patternicus.
5. Grandmother Smith refers to Patternicus as "my Patty."
6. Mr. and Mrs. Smith are becoming more and more frustrated with Patternicus's behavior.
7. The Smiths are considering selling Patternicus and have contacted someone who is interested.

The Disappearing Dog Dilemma (student-centered version) *(Cont'd.)*

8. Grandmother Smith has argued daily with Mr. and Mrs. Smith over the fact that Patternicus may be sold.

9. There have been daily deliveries of chicken to Grandmother Smith's house for the past week.

The kennel

1. The only footprints outside the kennel were a single set of dog prints, leading away from the building.

2. Kennel workers said that Patternicus was acting strangely before the disappearance, staring at an open window and barking.

Other

1. When the dog prints were run through the crime lab computer, they came back as a perfect match to Patternicus's records.

2. An empty bag of chicken treats was found on the curb outside the kennel.

3. A reward has been offered, but so far no one has responded.

Your analysis:

Who took Patternicus?	
Events that support your theory	Patterns that support your theory

Name _____ Date _____

Reflecting on Patterns

1. What have I learned about patterns?

2. What is still difficult or confusing for me about patterns?

3. What new questions do I have about patterns at the end of these lessons?

4. How can recognizing and using patterns help solve problems?

5. How can recognizing and using patterns help in other areas of my life?

Chapter 7

Measurement

Measurement occurs daily, in a variety of contexts and for a variety of reasons. It is virtually impossible to complete a day without engaging in a myriad of measurement-dependent moments. It is important, then, that education not only prepare students to engage in computations that permit on-time arrivals and provide enough paint to cover a specific amount of wall space but also to have a sense of time as well, so that "In a minute" has a real meaning, as do comparative phrases, and students can eventually ponder questions like, "Does bigger mean more mass?" The lessons in this chapter attempt to capture both the very formal nature of some classroom measurement work (seen in the grades 2/3 and 4/5 lessons) and the kind of embedded approach that maximizes the natural measurement opportunities inherent in everyday classroom activities (most evident in the grades K/1 lessons).

Measurement – Grades K/1

Teacher-Directed Lesson

In this lesson, as part of developing an understanding of the concept of measurement and the use of tools to measure, the teacher guides students through activities that relate to measuring length. After having outlines of their bodies traced and the images cut out by the teacher, students use Unifix cubes to measure parts of their bodies. They then move on to measuring items and objects in their classroom.

Lesson Outcomes	Students will use Unifix cubes to measure and report the length.
NCTM Standards	Measurement Standard:
	Understand measurable attributes of objects and the units, systems, and processes of measurement.
	Apply appropriate techniques, tools, and formulas to determine measurements.
Materials Needed	Butcher or other long roll of paper, Unifix cubes, chart paper, overhead or black/whiteboard
Time Required	1 hour, with other follow-up opportunities as desired or needed
Teaching Practices	Direct instruction
Assessment Purposes	Formative – all measuring activities

Teaching Guide

1. Trace an outline of each student on butcher paper. Cut out the images (students can personalize their images by coloring and adding clothing—especially if that supports other content curriculum outcomes—or these can remain more as paper-doll cutouts).

2. Provide students with ten Unifix cubes apiece. Leave another five to ten cubes at each table or in centrally located areas that students have access to.

3. Instruct students to connect ten Unifix cubes, making a ten-cube bar.

4. Show students how to use their bars and cubes to measure a piece of paper.

a. Students practice using the Unifix bar and extra cubes to measure a similar piece of paper so the teacher can observe how they are measuring and clarify confusions if they exist.

b. Students now use the Unifix bar, and extra cubes if needed, to measure body parts on their paper likeness (such as length of an arm, length of a leg, length of one of their feet, width of their head, waist, wrist).

5. Record all measurements on a posted chart.

6. In small groups, work with students who have difficulty.

7. Instruct students to use their Unifix bars and extra cubes to measure classroom items (such as height of the seat of their chair, length of the leg of their desk, the edges of a book, the width of a window or door). If necessary, they can make and use additional bars of ten.

8. Collect students' responses and use as evidence of learning.

Measurement – Grades K/1

Student-Centered Lesson

Designed to maximize student curiosity and thinking, this lesson is conceptualized as a series of inquiry-based experiences built around questions like "How big am I?" "How long does it take?" and "How heavy is it?" In exploring the questions, students are involved in experiences that move them to compare and order objects by size, travel the room to explore measuring distance, and see if they can stand the test of time. Additional questions and suggestions are included to enable teachers to customize measurement work to the reality of their own classroom.

Lesson Outcomes	Students will understand that size, time, and weight are measurable quantities and will explore measurement using nonstandard units. Students will develop a sense of comparative measures (bigger than, heavier than, and so on).
NCTM Standards	Measurement Standard:
	Understand measurable attributes of objects and the units, systems, and processes of measurement.
	Apply appropriate techniques, tools, and formulas to determine measurements.
	Problem-Solving Standard:
	Build new mathematical knowledge through problem solving.
	Solve problems that arise in mathematics and in other contexts.
	Apply and adapt a variety of appropriate strategies to solve problems.
	Monitor and reflect on the process of mathematical problem solving.
	Communication Standard:
	Communicate their mathematical thinking coherently and clearly to peers, teachers, and others.
	Use the language of mathematics to express mathematical ideas precisely.

Materials Needed	Butcher or other rolled paper, Unifix cubes, balance scales, objects for weighing
Time Required	15 to 30 minutes per exploration. Explorations can be done independent of one another, as paired events, or as center activities, as long as connections to measurement are made consistently.
Teaching Practices	Questioning, facilitation of performance-based learning experiences, small group and whole class work
Assessment Purposes	Formative – Observations during any of the steps of the explorations, especially measuring with Unifix cubes and responses to the exploration title questions
	Summative – Statements on back of cutouts; measuring self or classroom objects; does bigger mean more mass?

Teaching Guide

Select from the following questions and suggested activities, or use them as a springboard for others that better match the needs of your students and curriculum.

"How Big Am I?"

Preparation: Trace an outline of each student on butcher paper. Cut out images (students can personalize their images by coloring and adding clothing, especially if that supports other content curriculum outcomes—or these can remain more as paper-doll cutouts).

Inquiry Experience 1

1. Play "Is it bigger than a shoebox?"

    a. Teacher picks and tells a category (such as pets) and then thinks of an item in that category (such as turtle). Students must guess by asking questions about the item's relative size (such as "Is it bigger than a shoebox?" "Is it smaller than a truck?" "Is it smaller than a mouse?")

2. After playing the game, ask students to carry their cut out around the classroom and answer the question, "How big am I?" by comparing their paper person to items in the room (I am smaller than the door, I am bigger than my chair, and so on).

3. Document their findings on chart paper, then let students select two of their own statements to write on the back of their paper person.

Inquiry Experience 2

1. Provide students with ten Unifix cubes apiece. Leave another five to ten cubes at each table or in centrally located areas that students have access to.

2. Instruct students to connect ten Unifix cubes, making a ten-cube bar.

3. Model for students how to use their bars and cubes to measure a piece of paper and have them try this at their tables or desks, checking to be sure that they are clear about what to do.

4. Give students their paper image cutouts and tell them that now they are going to use these to help them find more to say about the question, "How big am I?"

   a. Students now use their paper cutout and the Unifix bar (and extra cubes if needed) to measure their overall height as well as the length of an arm, leg, or one of their feet; the width of their head, waist, wrist; and so on.

   b. In small groups, work with students who may be having difficulty.

5. Record all measurements next to each child's name on a posted chart. Transfer these to the back of the paper cutout, and save this to be used during the year and at the end of the year to show how much the student has grown.

An extension of "How big am I?" is "How big is it?" For this, students measure objects and items in the classroom using their Unifix-cube bar. They can also create their own Unifix-cube ruler by tracing the bar and marking where each cube joins the next, then cutting it out. This Unifix ruler can go home with students so they can continue exploring "How big is it?" in their homes and neighborhoods.

"How Long Does It Take?"

1. Ask students "How long is a minute?" and allow them to discuss this.

2. Tell students that you want them to wait a minute, then raise their hands (have them put their heads down or close their eyes so that they aren't influenced by the raising of hands around them).

Document how many seconds have actually passed when hands are raised, then share this with students.

3. Now tell students that they have to be completely still and silent for one minute. Time them and tell them when a minute is up. How long did it feel like?

   This can be extended by asking students to note when they think five minutes have passed.

4. Predict "How long it will take for . . . ?"
   - An ice cube to melt
   - The perfume smell to reach the other side of the classroom
   - Two people to play a game of tic-tac-toe
   - The class to be packed up to go home
   - Us to finish this activity

5. Measure the actual time and compare it to the predictions. How close were we?

   These questions can be asked all year long, as they relate to a variety of contexts, in and out of the classroom.

"What Has More Mass?"

1. Set up several stations with balance scales, then model for students how to use the scales to compare mass.

   a. Students hold a paper clip and a wooden block, and the teacher asks which is heavier.

   b. Students put the paper clip on one side of the balance scale and the block on the other.

   c. Ask them what they see (the side with the block is lower than the paper-clip side).

   d. Repeat steps a through c with two different objects.

   e. Finally, have students just use the scale, and ask them which is heavier based on what they see the scale doing.

      Explain that when something has more mass, it weighs more.

2. Pose the question "What has more mass?"

   a. Students use the balance scale to weigh different objects on each side of the scale (for instance, a teddy-bear counter on one side and a crayon on the other side). It is important that they use only one of each item during this part of the activity.

   b. Students continue to look at the question of more mass by examining sets of like objects (such as a set of two teddy-bear counters and a set of three teddy-bear counters).

c. Students further deepen their thinking about mass by considering the question, "Does bigger mean more mass?" and examining a golf ball and an inflated balloon to help them have the conversation.

Here are some other questions that can be explored:

- "How much can it hold?" (measuring volume or capacity)
- "How much space is there?" (measuring area)
- "How far is it?" (measuring distance)
- "How cold/warm is it?" (measuring temperature)

Measurement – Grades 2/3

Teacher-Directed Lesson

The teacher leads the class through a lesson on standard and non-standard measurement. Students actively engage in structured measurement activities.

Lesson Outcomes	Students will understand the differences between, and advantages of, standard and nonstandard measurement; students will measure accurately using a ruler.
NCTM Standards	Measurement Standard:
	Understand measurable attributes of objects and the units, systems, and processes of measurement.
	Apply appropriate techniques, tools, and formulas to determine measurements.
Materials Needed	Unifix cubes, teddy-bear counters or blocks, chart paper (or overhead or black/whiteboard), shoeboxes, balance scales, chalkboard erasers, class set of rulers, student worksheets:
	7.1: How Big Is It?
	7.2: A Different Way to Measure
	7.3: Measuring with a Ruler
Time Required	3 hours 30 minutes
Teaching Practices	Explanation/lecture, student performance, discovery
Assessment Purposes	Formative – Small- and large-group discussions, Worksheets 7.1 and 7.2
	Summative – Worksheet 7.3

Teaching Guide

1. Explain that we measure things in order to see how long they are (their *length*) or how much they weigh (their *weight*) or how much they can hold (their *capacity*).

2. Tell students that they will be measuring different things in different ways, then they will all share their responses.

3. Distribute *Student Worksheet 7.1: How Big Is It?*

   a. Students follow the instructions in the worksheet, using nonstandard units of measure to determine the length of their desk or table and the classroom wall, the volume of a shoebox, and the weight of an eraser.

   b. Students compare their measurements, noting differences.

4. After all students have had an opportunity to measure each item, they share their answers as a class. Using an overhead projector, blackboard, or chart paper, the teacher documents student responses.

5. Ask students which measurements were the most different? Which were the most alike?

6. Discuss with students the fact that there are differences in the measurements, explaining that the things they used to measure were different sizes, so the measurements would be different. Explain that even though everyone used their hands or their feet, the measure would probably be different because each child's hands and feet are a different size than those of the other children.

   If necessary, have students compare the sizes of their hands and feet, asking the question, "Do two of your hands measure the same as two of your partner's hands?" This can then be used as a way to help students understand why, when they used hands and feet to measure, the measurements were often different.

7. Explain capacity and weight measures in a similar manner, showing students that their answers depend on the items they selected to use as measurement units. Help students understand that when they used the same measurement unit (like Unifix cubes), the measurements came out more nearly the same than when they used completely different units (such as using anything they wanted to balance the scale).

8. Ask students, "What would help us to all get the same measurement?"

   Sample student responses:

   "If we all used the same things to measure with."

   "If the things we used to measure with were all the same size."

   "If we all had the same thing to measure with."

9. Tell students that this is why we have standard units of measure like inches and centimeters. Even though we measure in "feet," the size of the foot that we use is always exactly the same (twelve inches).

10. Show how to use a ruler to measure length.

11. Distribute rulers and *Student Worksheet 7.2: A Different Way to Measure.* Have students work in pairs.

    a. Students remeasure their desks and the wall using a ruler.

    b. They report out their measurements while the teacher documents them—or students themselves write them on a piece of chart paper.

    c. Once all pairs have reported or written their new measurements, the class compares them to see how close they are to one another.

12. Ask which was better at giving the same or nearly the same measurement: student hands and feet, or the ruler. When students respond that it was the ruler, explain that this is the difference between standard and nonstandard measurement. Standard measurement means that the same measuring tool and the same units of measurement are used (ruler and inches or feet or meters). Nonstandard measurement means that different tools and units could be used (student hands and feet).

13. Conclude the lesson by reinforcing how measurements taken with a ruler, using the standard measure of a foot, are much more consistent than the measurements using student hands and feet.

14. Distribute *Student Worksheet 7.3: Measuring with a Ruler.* Have students complete the measurements required individually, and collect this to assess for understanding and evidence of learning.

Name _____   Date _____

How Big Is It?

You will be measuring different things in different ways, then sharing your responses.

1. Use your hand to measure your desk or table. How many hands long is your desk or table?

2. Measure the wall of our classroom with your feet. How many of your feet long is the wall?

3. Use Unifix cubes, teddy-bear counters, or blocks to measure how much a shoebox holds. What is the capacity of the shoebox?

4. Use the balance scale to measure the weight of a chalkboard eraser. You may use any objects you have to balance the scale. How much does the eraser weigh?

Name _____   Date _____

Worksheet 7.2

A Different Way to Measure

With your partner, use your ruler to measure your desk and the wall. Write your measurements in the space below.

Length of my desk:

Length of the wall:

Worksheet 7.3

Measuring with a Ruler

Use a ruler for the following measurements. Be sure to measure carefully.

- The width of the classroom door

- The longest side of your notebook

- How far the seat of your classroom chair is from the ground

- A classmate's height

- The width of a window or closet

Measurement – Grades 2/3

Student-Centered Lesson

Students will investigate different forms of measurement as they focus on the question, "How can we measure well?"

Lesson Outcomes	Students will understand the purposes of measurement, the relationship between different types of measurement and tools of measurement; students will measure accurately using a ruler and other appropriate measurement tools (for example, scale, calendar, clock).
NCTM Standards	Measurement Standard:
	Understand measurable attributes of objects and the units, systems, and processes of measurement.
	Apply appropriate techniques, tools, and formulas to determine measurements.
	Communication Standard:
	Organize and consolidate their mathematical thinking through communication.
	Communicate their mathematical thinking coherently and clearly to peers, teachers, and others.
	Analyze and evaluate the mathematical thinking and strategies of others.
	Use the language of mathematics to express mathematical ideas precisely.
Materials Needed	Overhead or chart paper; copies of lesson charts for students; materials for research centers (books, pictures, videos); measurement tools (ruler, clock, calendar, scale); worksheets:
	7.4: How Can We Measure Well?
	7.5: Measuring
	7.6: Measuring Length
	7.7: Measuring Capacity

7.8: Measuring Time

7.9: Measuring Weight

7.10: Combined Group Responses

Time Required	5 to 15 days
Teaching Practices	Use of a question as an organizing center, research, student decision making, establishing and applying criteria, cooperative learning, reflection, authentic assessment
Assessment Purposes	Diagnostic – Initial, individual student responses to "How Can We Measure Well?" Formative – Worksheets 7.5, 7.6, 7.7, 7.8, and 7.9 Summative – Displays and sharing with other students

Teaching Guide

1. Pose the question, "How can we measure well?" Distribute *Student Worksheet 7.4: How Can We Measure Well?* for students to use to document their thinking.

   a. Students respond to the question in writing.

   b. In small groups of three or four, students share their responses.

   c. Using a different color pencil or pen, students revise their responses based on what they have heard in their small groups.

   d. Students share their responses as a whole class, generating a list that the teacher documents on overhead or chart paper.

2. Distribute *Student Worksheet 7.5: Measuring*. In pairs or triads, have students think about and answer the following questions:

   • What do we measure?

   • How do we measure it?

   • What are the tools that we use to measure?

3. Chart responses and facilitate a conversation about these questions as groups report out.

   Sample responses:

   What do we measure? Length, weight, time, capacity . . .

   How do we measure it? Inches, feet, meters, pounds, tons, minutes, years . . .

   What are the tools that we use to measure? Ruler, scale, clock, calendar . . .

4. Students work in small groups, formed around the answers to "What do we measure?" There might be a *time* group and a *length* group and a *capacity* group. Distribute Student Worksheets 7.6, 7.7, and 7.8 as appropriate.

   It may be necessary to adjust the group names at the top of the worksheets to match the groups that the class forms.

5. Form new groups by including one student from each of the previous groups (time, length, capacity, weight, and so on). Distribute *Student Worksheet 7.10: Combined Group Responses*. Members share information from their completed charts to create a single combined chart. Groups can add to any category.

6. Students create displays of the categories that they have discussed (length, time, capacity, weight, and so on), including the tools and examples or pictures of the kinds of things that would be measured using those tools. They share their displays with younger students, teaching them about what and how we measure.

Name _____ Date _____

How Can We Measure Well?

1. In the following space, write your ideas about how we can measure well.

2. Share your response with three or four of your classmates.

3. Once you've shared your responses and listened carefully to what your partners have said, use a different color pencil or pen to revise your responses based on what you have heard.

Measuring

Think about and answer the following questions. You may work in pairs or triads.

- What do we measure?

- How do we measure it?

- What are the tools that we use to measure?

Measuring Length

Group: _____

Members: _____

_____

What kinds of things do we measure the length of? Why?	How do we measure its length? (unit)	What tool(s) do we use to measure its length?

Measuring Capacity

Group: _____

Members: _____

_____

What kinds of things are important to know the capacity of? Why?	How do we measure capacity? (unit)	What tool(s) do we use to measure capacity?

Measuring Time

Group: _____

Members: _____

_____

What kinds of things are important to know the time of? Why?	How do we measure time? (unit)	What tool(s) do we use to measure time?

Measuring Weight

Group: _____

Members: _____

_____

What kinds of things are important to know the weight of? Why?	How do we measure weight? (unit)	What tool(s) do we use to measure weight?

Combined Group Responses

Members: _____

_____

_____

	What kinds of things do we measure?	Why do we measure things?	What units of measure do we use?	What tools do we use to measure?
Length				
Capacity				
Time				
Weight				

Measurement – Grades 4/5

Teacher-Directed Lesson

In this lesson, students create a scale drawing of their classroom.

Lesson Outcomes	Students will measure accurately and draw to a specified or chosen scale.
NCTM Standards	Measurement Standard:
	Understand measurable attributes of objects and the units, systems, and processes of measurement.
	Apply appropriate techniques, tools, and formulas to determine measurements.
Materials Needed	Books, blueprints, scale models, scale drawings, graph paper
Time Required	1 hour
Teaching Practices	Guided practice, small-group targeted instruction
Assessment Purposes	Formative – Measurement opportunities before the scale drawing is begun; measuring the perimeter
	Summative – Final performance task: the scale drawing

Teaching Guide

1. Show students samples of blueprints or plans for buildings, model airplanes or ships, or any of David Macaulay's books about structures to begin a conversation about drawing to scale. Ask students why it might be necessary to be able to create something "to scale."

2. Distribute graph paper and tell students that they are going to make a scale drawing of the classroom, including furniture.

3. Either assign them a scale or allow them to choose their own.

4. Before allowing them to begin their scale drawings, provide them with opportunities to measure specific objects (such as a desk, window, book) to check for accuracy. At this point, if there are students who are having difficulty taking accurate measurements, work with them in small-group targeted instruction.

5. Instruct students to measure and draw the perimeter of the room. Check for accuracy (this is another point at which small-group instruction could be helpful).

6. If there are specific items or areas that you want represented, be sure to let the students know ahead of time.

7. Students measure and draw a scale drawing of the classroom.

8. Drawings are assessed for accuracy and neatness, then posted around the classroom or on the bulletin board.

Measurement – Grades 4/5

Student-Centered Lesson

Students think and behave as interior designers, reinventing their classroom space. They create and submit scale drawings as the main part of their proposal for redesign. Proposals are evaluated, and some are selected for implementation on a monthly basis.

Lesson Outcomes	Students will measure accurately and draw to a specified or chosen scale. Students will apply measurement tools and skills to an authentic context.
NCTM Standards	Measurement Standard:
	Understand measurable attributes of objects and the units, systems, and processes of measurement.
	Apply appropriate techniques, tools, and formulas to determine measurements.
	Communication Standard:
	Communicate their mathematical thinking coherently and clearly to peers, teachers, and others.
	Analyze and evaluate the mathematical thinking and strategies of others.
	Connections Standard:
	Recognize and apply mathematics in contexts outside of mathematics.
Materials Needed	Books, blueprints, scale models, scale drawings, graph paper, Worksheet 7.11
Time Required	45 minutes for instruction; approximately 1 week for redesign proposals and presentations
Teaching Practices	Guided practice, small-group targeted instruction, facilitation of student independent work, authentic assessment, design and use of rubric and/or checklist or other form of explicit criteria

Assessment Purposes   Diagnostic – Worksheet 7.11

Formative – Feedback from self and peers

Summative – Final performance task—the scale drawing and proposal

Teaching Guide

1. Show students samples of blueprints or plans for buildings, model airplanes or ships, or any of David Macaulay's books about structures to begin a conversation about drawing to scale. Ask students when it might be important to be able to create something "to scale."

   If this is the first time that scale drawing is introduced, then the teacher should distribute graph paper and rulers, then have students draw the following using the scales provided:

   Using the scale 0.5 inch = 1 foot, draw a rectangular room whose real dimensions are 15 feet by 12 feet.

   Using the scale 1 inch = 1 foot, draw the perimeter of a garden plot that is 6 feet by 9 feet.

   If students require more practice, continue to provide situations and scales for them to work with. You may want to pair them for some drawings and let them work individually on others. Once students are comfortable drawing to scale, proceed to the rest of the lesson.

2. Explain to students that you are interested in changing the classroom, and you want to make sure that the space is used well, that it is comfortable and supports learning—but you aren't sure what the best arrangement might be, so you are going to turn the problem over to them, and the students will be the interior designers.

3. Each student will submit a proposal for redesigning the classroom space. Part of every proposal will be a scale drawing as well as a short presentation describing the new design and explaining the changes.

   This is a good opportunity to engage students in a conversation about what a quality redesign proposal would include (for example, at least one accurate, neat, and clear scale drawing of the proposed design; changes that are doable, require little or no money, and are good for learning, are comfortable, and are different from the current configuration).

4. Distribute graph paper and the *Pre-test/Diagnostic Worksheet 7.11: How Accurately Do I Measure?*

5. Use data from the diagnostic to identify any students who have difficulty taking accurate measurements or using a scale; work with them, individually or in small groups, to clarify questions and solidify skills.

6. Students research classroom designs, looking on the Internet and in books, if available. They make some decisions about the strengths and needs of the current classroom design, then brainstorm changes.

7. Students peer-review their ideas, giving feedback and asking questions using the components and criteria that they listed earlier. Then students use the feedback to revise their ideas.

8. On their graph paper, students create a scale drawing of their proposed design (if desirable, they can use the graph paper from the pre-test, as the perimeter of the room is already there).

9. They prepare their proposal presentation, making sure that all measurements and calculations (if needed) are precise and correct, and looking at the required components and criteria to self-assess.

10. Before finishing, students have the opportunity to get feedback one more time from a member of the class and to use that feedback as another revision opportunity.

11. Students present their proposals to the class. The teacher and a panel use the components and quality criteria as measures to evaluate the proposals. Those proposals that best meet the criteria become the "finalists," from which the designs that will actually be used are selected.

    Some options:

    • The "panel" can be composed of a combination of older students, architects or interior designers from the community, and teachers and administrators. It is important that they understand the criteria to which the students are being held so that they can apply it well.

    • If there are many designs that meet all criteria, enough can be selected to change the room once during every remaining month of the school year.

Name _____   Date _____

How Accurately Do I Measure?

1. Measure the perimeter of the room.

    P = _____

2. Measure the width of one window.

    W = _____

3. Using a scale of 0.5 inch = 1 foot, draw the perimeter of our classroom on your graph paper.

What are some ideas that you have for redesigning our classroom?

Chapter 8

Money

Being able to understand and manipulate money is a life skill of unquestionable importance. Like it or not, have it or not—whether plastic or paper, electronic or coin—money and all that it enables is at the heart of much that everyday life encompasses, from philanthropy to conflict, at home and in the world.

This chapter focuses on helping students grasp the meaning of money, from a concrete perspective in primary work to the more abstract realm of budgets and funding wish lists, for older children. The lessons in this chapter exemplify three different design choices. K/1 students are involved in concrete activities of recognition, manipulation, and the basics of using money to purchase items, while in grades 2/3, literature is the base of the lesson. The 4/5 lesson is designed as an ongoing performance task. Each lesson ends in a simulation, providing an opportunity for students to put into practice what they have learned.

Money – Grades K/1

Teacher-Directed Lesson

In this lesson, the teacher guides learning through direct instruction, modeling, and guided practice. Student independent work is limited to completing a series of defined tasks.

Lesson Outcome	Students will be able to recognize the penny and nickel, count by ones and skip count by two, state the cent value of a penny and a nickel, and recognize five pennies as the equivalent of a nickel.
NCTM Standards	Numbers and Operations:
	Understand numbers, ways of representing numbers, relationships among numbers, and number systems.
	Algebra Standard:
	Understand patterns, relations, and functions.
	Use mathematical models to represent and understand quantitative relationships.
Materials Needed	Black/whiteboard or overhead, coins, pencils, student worksheets:
	8.1: Penny
	8.2: Nickel
	8.3: How Much Do I Have?
Time Required	1 hour
Teaching Practices	Modeling, guided practice, independent practice
Assessment Purposes	Formative – Worksheets 8.1 and 8.2

Teaching Guide

1. Show a large picture of a penny and tell students, "This is a penny." Students say "penny." Repeat this, either by pointing and saying "penny" while students repeat, or having a student come up to the picture, point to it, and name it, then the other students repeat.

2. Post four more pictures of pennies, so that there are five in all. Model the way that they are posted so that it supports the layout of Worksheet 8.1 (coins lined up in a row).

3. Distribute *Student Worksheet 8.1: Penny* and provide pairs of students with several pennies and nickels.

4. Have students separate out the pennies in their pile of coins.

   a. On their worksheets, students place their five pennies on a circle. To reinforce counting, have students count aloud as they place the coins in each circle (one penny, two pennies, three pennies . . .).

   b. Pointing to each penny, they state its name.

5. Tell students that every penny has two different sides. Have them turn their pennies over and look at the other side.

6. Have students make pencil or crayon rubbings of the two sides of the penny (use the "My Penny" space on the worksheet, or a separate piece of tracing or other thin paper). Assist children for whom this task is difficult.

7. Distribute *Student Worksheet 8.2: Nickel.*

8. Repeat steps 1–6 with nickels.

9. Distribute *Student Worksheet 8.3: How Much Do I Have?*

10. Explain to students that a penny equals one cent.

    a. Show students a penny with "____ = 1 cent" written next to it.

    b. Show two pennies with "= cents" written next to it, and ask students to tell what number belongs in the blank.

    c. Students look at their worksheets and respond to questions 1–5, using their real coins, counting the pennies, and writing the amounts.

    Explain to students that a nickel equals five cents.

    Show students a nickel with "____ = 5 cents" written next to it.

    Show two nickels with "____ = cents" written next to it, and ask students to tell what number belongs in the blank.

    This is a good time to introduce or reinforce skip-counting by five.

    d. Students look at their worksheets and respond to questions 6–10, using their real coins, skip-counting, and writing the amounts.

Money   **185**

11. Show students the picture of five pennies and one nickel.

    a. Ask students to tell how many cents five pennies equal. Write the number 5. Keep this posted.

    b. Point to the nickel, and ask students how many cents a nickel equals. Write the number 5. Post this next to or underneath the pennies.

    c. Ask students what they notice about five pennies and a nickel (both are the same/both are equal/both equal 5 cents). If no one says that five pennies equal one nickel, use that statement to summarize what was said.

    d. In pairs, have students take turns making change, giving each other five pennies for a nickel or a nickel for five pennies.

    e. In their worksheet, students should draw a picture of the statement, "Five pennies equal one nickel."

12. Distribute *Student Worksheet 8.4: Do I Have Enough?*

    a. Tell students that you pay for things with money, and it is important to know if you have enough to buy what you want.

    b. Students complete the worksheet and have the teacher check their answers. For each answer, they are awarded a penny.

    c. Once students have five pennies, they can trade them for a nickel and use it to buy stickers or some other small reward.

    d. Ideally, all students will get five pennies in a reasonable amount of time. It may be necessary to differentiate at this point to facilitate this. Students who have difficulty or do not get all five correct can receive small-group or individual instruction or review, and then can try the problems over. For students who still need guidance, the teacher can sit with them and provide more guidance as students try the problems over.

13. Collect Worksheet 8.4, along with any notes you may have made based on how many times the problems were tried, or on how much or what kind of intervention was needed for the student to meet with success.

Penny

Place one penny on each circle.

◯ ◯ ◯ ◯ ◯

My Penny

Name _____   Date _____

Nickel

Place one nickel on each circle.

○   ○   ○   ○   ○

My Nickel

Name _____   Date _____

How Much Do I Have?

How many pennies, how many cents?

1)   = _____ CENTS

HOW MANY PENNIES? _____

2)   = _____ CENTS

HOW MANY PENNIES? _____

3)   = _____ CENTS

HOW MANY PENNIES? _____

4) = _____ CENTS

HOW MANY PENNIES? _____

5)   = _____ CENTS

HOW MANY PENNIES? _____

Name _____ Date _____

How Much Do I Have? *(Cont'd.)*

How many nickels, how many cents?

6)   = _____ CENTS

   HOW MANY NICKELS? _____

7)   = _____ CENTS

   HOW MANY NICKELS? _____

8)   = _____ CENTS

   HOW MANY NICKELS? _____

9)   = _____ CENTS

   HOW MANY NICKELS? _____

10)   = _____ CENTS

   HOW MANY NICKELS? _____

Name _____   Date _____

Do I Have Enough?

Shopping list	5 pennies or 1 nickel?	
Pencil 5 cents	5 PENNIES	1 NICKEL
Plastic teddy bear 5 cents	5 PENNIES	1 NICKEL
Apple 5 cents	5 PENNIES	1 NICKEL
Crayons 5 cents	5 PENNIES	1 NICKEL
Toy truck 5 cents	5 PENNIES	1 NICKEL

Money – Grades K/1

Student-Centered Lesson

In this lesson, the teacher guides student learning through a series of experiential learning opportunities that engage students in recognizing and using pennies and nickels.

Lesson Outcome	Students will be able to recognize, name, and tell the amount of a penny and a nickel. They will count by ones and skip count by ^{2} and will begin to explore the relationship between the respective values of the two coins as well as the concept of using money to purchase goods.
NCTM Standards	Number and Operations:
	Understand numbers, ways of representing numbers, relationships among numbers, and number systems.
	Algebra Standard:
	Understand patterns, relations, and functions.
	Use mathematical models to represent and understand quantitative relationships.
Materials Needed	Black/whiteboard or overhead; five pennies and five nickels per student; plastic bags with students' names, items for purchase (plastic teddy bears, toy trucks, pencils, crayons, apples, or the equivalent); student worksheets:
	8.5: Which Coin Is Which? (Exploration 1)
	8.6: How Many? (Exploration 2)
	8.7: What Are They Worth? (Exploration 3)
	8.7A: Change Form (Exploration 3)
	8.8: Student Reflection
Time Required	Exploration 1 – 1.5 hours; Exploration 2 – 45 minutes; Exploration 3 – 1 hour; Interview/Reflection – 45 minutes
Teaching Practices	Questioning, facilitating student inquiry, guided practice, independent practice, small-group work, student reflection
Assessment Purposes	Formative – All small-group and class conversations, teacher observations of students' independent and small-group work, worksheets 8.5, 8.6, 8.7, and 8.7A
	Summative – Worksheet 8.8

Teaching Guide

Each exploration is a self-contained activity that is debriefed in small groups or as a whole class. Though they are meant to occur in sequence, explorations can be broken into component parts and do not have to be implemented all at once.

Exploration 1 – Which Coin Is Which?

Students will recognize and differentiate between pennies and nickels.

1. Post a large picture of a penny and one of a nickel. At the same time, provide pairs of students with five or six of each of the actual coins. Give students time in pairs to examine the coins and visit the pictures, looking carefully for what is the same and what is different.

    a. Ask students to share what they noticed about the penny (point to the penny) and the nickel (point to the nickel), documenting their observations on chart paper (these can be written as two lists—Same and Different—or as a Venn diagram). Prompt students to look again, if necessary.

    b. Consistently refer to the two coins by their names as part of the conversation. At the end of the discussion, if no student has noted that the names are different, say the name of each one clearly, and prompt students by asking what is different about what each is called.

2. Distribute *Student Worksheet 8.5: Which Coin Is Which?*

    a. Students use their pennies and nickels to create two different repeating patterns (such as penny, penny, nickel, penny, penny, nickel). They copy these patterns onto their worksheets.

    b. Individually, students select one penny and one nickel from their shared group of coins and make pencil rubbings of both sides of each.

    c. Cut rubbings out and tape them to two clear plastic bags that have been prelabeled with student names, so that the bags are now also labeled with either the front and back of a penny or the front and back of a nickel (each student should have one bag with his or her name for pennies and one for nickels).

3. Distribute ten more pennies and ten more nickels to each pair of students (they now each have ten of each coin).

    a. Students separate pennies from nickels.

    b. Students count five pennies and put them in their penny bag, and count five nickels and put those in their nickel bag.

c. They count the remaining coins and return them to the teacher.

d. Students use their coin bags for the two Explorations that follow.

Exploration 2: How Many?
Students will count coins by ones and will skip count by two.

Divide students into groups of three. Distribute *Student Worksheet 8.6: How Many?*

a. Have students take their coins out of the bags and combine all of them into one group.

b. Students take turns counting the coins, then line them up in a row. How many pennies do they have? (15) How many nickels? (15) How many coins, total? (30)

If students cannot yet independently count items to 30 reliably, adjust the number of coins to an amount that seems reasonable.

c. Students count the 30 coins individually as quickly as they can, without sacrificing accuracy, then count them by twos.

d. In their groups, one child should count the coins by ones and one by twos, while the third should be the recorder, keeping track of which method of counting goes faster. Each student should have the opportunity to participate in each role twice, so there are a total of six trials.

e. Groups looking at the data that they have collected should report out which they believe is a faster counting method: by ones or by twos.

f. Students compare the results of all of the groups and state a hypothesis about the connection between speed and counting by ones and twos.

Exploration 3 – What Are They Worth?
1. Ask students what people use money for (to pay for things, to buy things, to go to the store, and so on). Tell them that they are about to use their money to buy things for themselves.

2. Distribute *Student Worksheet 8.7: What Are They Worth?* and tell students that this is their shopping list. They must buy all five things, but they can decide which coins to use when they make a purchase.

a. Before going to the store, students look at their shopping list and decide how to pay for the first item, circling either five pennies or one nickel.

b. Students take the amount of money out of their bags and take turns going to the "store" to purchase the items on the list.

c. This process is repeated for all five items. If students do not have the coins that they want to use, they can take one turn to stop at the "bank" on their way to the store, and change money by filling out *Student Worksheet 8.7A: Change Form*, indicating the coins they have and the coins they want. (Both the bank and the store are operated by the teacher.)

d. After all students have purchased the items on their list, they look at the money left in their bags. (There should be the equivalent of five cents, in either pennies or a nickel.)

3. Distribute *Student Worksheet 8.8: Student Reflection*.

a. In pairs, students interview one another about how they paid for the different items. Together, they respond to questions about why they think that they could pay with either five pennies or a nickel, why they could swap five pennies for a nickel or a nickel for five pennies at the bank, and why they have either a nickel or five pennies left at the end.

b. Groups report out to the whole class. If students do not mention anything about five pennies and a nickel being the same or equal, ask more questions that will help them think about this.

If students cannot do this independently, then the teacher can either interview each student separately and keep the responses as evidence of learning, or ask the questions aloud and have students discuss them in pairs, reporting out to the class after each question, while the teacher documents their responses.

c. Students respond to the reflective prompt:

What have you learned about pennies and nickels?

Worksheet 8.5

Which Coin Is Which?
(Exploration 1)

My Patterns

1.

2.

Penny Rubbing

Nickel Rubbing

Name _____ Date _____

How Many? (Exploration 2)

Trial Data

For each trial, place an X in the box to show which way of counting was faster.

TRIAL	Count by 1	Skip count by 2
1		
2		
3		
4		
5		
6		

Which is fastest?

Name _____ Date _____

What Are They Worth?
(Exploration 3)

Going Shopping

Shopping list	5 pennies or 1 nickel?
Pencil 5 cents	5 PENNIES 1 NICKEL
Plastic teddy bear 5 cents	5 PENNIES 1 NICKEL
Apple 5 cents	5 PENNIES 1 NICKEL
Crayons 5 cents	5 PENNIES 1 NICKEL
Toy truck 5 cents	5 PENNIES 1 NICKEL

Name _____   Date _____

Change Form
(Exploration 3)

Change Form

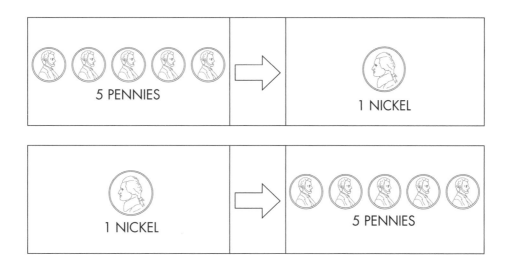

5 PENNIES → 1 NICKEL

1 NICKEL → 5 PENNIES

Name _____ Date _____

Student Reflection

Interview questions:

1. How much did each item on your shopping list cost?

   _____

2. How did you decide whether to pay with five pennies or a nickel?

   _____

   _____

   _____

   _____

3. Why do you think you had a choice of how to pay?

   _____

   _____

   _____

   _____

4. Did you change money at the bank? Why or why not?

   _____

   _____

   _____

   _____

Student Reflection *(Cont'd.)*

5. How much money do you have left?

_____

6. Do you have five pennies or a nickel left?

_____

7. How many cents are five pennies equal to?

_____

8. How many cents is a nickel equal to?

_____

Reflection

What are some very important things that you have learned about money?

Money – Grades 2/3

Teacher-Directed Lesson

Beginning with a review of the four coins, this lesson—adapted from a lesson by West Islip, New York teacher Celia Gottlieb—uses a children's story as the context of a series of specific activities related to adding and subtracting pennies, nickels, dimes, and quarters. The lesson supports students' recognition and calculation skills by involving them in a process of earning, saving, and spending money.

Lesson Outcome	Students will be able to add and subtract different amounts of money.
NCTM Standards	Number and Operations:
	Understand meanings of operations and how they relate to one another.
	Compute fluently and make reasonable estimates.
Materials Needed	Black/whiteboard or overhead; large pictures of a penny and a nickel; pennies, dimes, nickels, and quarters for Money Grab Bag (if used); the book *You Can't Buy a Dinosaur with a Dime* by Harriet Ziefert; eight dinosaurs that can be purchased; coins for allowance and chores per story; student worksheets:
	8.9: Reviewing Where We Are and What We Have
	8.10: You Can't Buy a Dinosaur with a Dime
Time Required	1 hour (plus 30 minutes for Money Grab Bag, if used)
Teaching Practices	Direct instruction, reading aloud
Assessment Purposes	Diagnostic – Assessing prior knowledge
	Formative – Student responses during review, Money Grab Bag (if used), group responses during reading aloud and activity
	Summative – Worksheet 8.10 (if completed individually)

Teaching Guide

1. Pose the questions "How much money is enough? How do you know?" and give students time to think about and discuss them.

Accessing Prior Knowledge

2. Show large pictures of a penny, nickel, dime, and quarter.

   a. Ask students to tell the name of each as it is held up.

   b. Do this several times, until students seem fairly comfortable with the naming.

3. Post the pictures of the coins (penny, nickel, dime, quarter).

   a. Ask the value of each.

   b. Make a game of the guessing and repeat this until students are comfortable.

   c. Have students write the values of the coins underneath each picture.

For students still having difficulty correctly naming or associating value with coins, this would be a good time to review or reteach, individually or in small groups.

4. Show students how to add the values of the coins.

   If students have already learned this, a quick review might be enough, but if more is needed, or as an ongoing review, one option is to create a game, like Money Grab Bag:

   a. Set up bags with several pennies, dimes, nickels, and quarters.

   b. Three or four students can work with one bag.

   c. Students take turns picking two coins out of the bag.

   d. The student draws the coins selected, labeling them with their names and with their values inside or underneath, determines their total value by adding, and returns the coins to the bag.

   e. Other students in the group check the player's work and help when needed.

   f. Students complete at least three rounds.

Students needing more targeted intervention may benefit from work with picture representations of coins—or real with ones. If real coins or realistic pictures of coins are not available circles with the value written in the middle. Create addition problems with the pictures or coins, and write the real values underneath them. This will allow students to see the problem as a basic computation (such as 10 + 10 = 20). With pictures that are less realistic, like the circles with coin amounts inside, the appearance of the problem is slightly different, but the basic computation is the same.

5. Form coin groups using the four coins just reviewed.
   a. Have students select a penny, nickel, dime, or quarter from a bag.
   b. A student states the coin and its value, then goes to stand by the picture of that coin.
   c. Once all students are part of a group, send groups to sit at tables. Split large groups into groups of three or four students.

6. Distribute *Student Worksheet 8.9: Reviewing Where We Are and What We Have* and bags of coins for activities.

7. Explain that students will complete the activities with their group while they listen to the story.

8. Distribute *Student Worksheet 8.10: You Can't Buy a Dinosaur with a Dime*.

9. Read the story *You Can't Buy a Dinosaur with a Dime* by Harriet Ziefert, pausing as appropriate to give students time to work through each part of the activity.
   a. Students then listen to the story and complete identification or computation activities, acting out parts of the story as they hand over money to buy a dinosaur and add money from the "chores" they have done.
   b. Students who have difficulty with the activities can be paired with other students in their group or can be regrouped in a way that allows for more direct support from the teacher.

10. To assess what has been learned, Worksheet 8.10 can be collected as evidence of student learning.

Reviewing Where We Are
and What We Have

1. Which coin group do you belong to? _____

2. What is the value of this coin? _____

3. How much money are you starting off with in your bag? _____

4. Of the money that you have, how many coins are:

   Nickels   _____

   Dimes   _____

   Quarters   _____

Directions for the rest of the lesson:

1. Listen carefully to the story.

2. Complete the questions on the activity sheet that you will be given.

3. You can work on the questions with your group, but you must fill out your own activity sheet.

Name _____ Date _____

You Can't Buy a Dinosaur with a Dime

1. Pete buys a dinosaur. Fill in the missing prices:

   Price of dinosaur = _____

   Sales tax = _____

   Total cost = _____

2. What coins in your bag can you use to pay for the dinosaur?
   (Hint: Make sure the total value of your coins equals the total value of the dinosaur plus tax.)

   Pennies   _____

   Nickels   _____

   Dimes   _____

   Quarters   _____

   Have one member from your group come up to buy the dinosaur—do not forget to bring the tax, too!

3. How much money do you have left in your bank? _____

4. Pete is upset about having an empty bank. With your group, write down three ways you can earn money.

5. When everyone in your group is done, share your answers. You can add to or change your answers based on what others say.

6. Pete earns money by doing chores.

   a. Have one group member come up and get the money that Pete earned.

   b. How many quarters did your group receive? _____

You Can't Buy a Dinosaur with a Dime *(Cont'd.)*

c. What is the value of this money? _____

d. In all, how much money do you have in your bank? _____

7. Pete earns allowance. Have one group member come up to receive Pete's allowance.

   a. How many coins did Pete earn for allowance?

   Nickels   _____

   Dimes   _____

   Quarters   _____

   b. What is the total value of Pete's allowance? _____

8. From his chores and allowance, how much money did Pete earn?

   Chores =   _____

   Allowance = _____

   Total =   _____

9. What is the value of the money in Pete's bank? _____

10. Pete finds a dime and sells six baseball cards for five cents each. What is the total value of the money that Pete received? _____

11. Have one group member come up and take the money Pete received.

12. Pete buys another dinosaur. Including tax, how much did Pete spend?

   Dinosaur =   _____

   Tax =   _____

   Total =   _____

13. Have one member from your group bring up the money Pete spent on his dinosaur.

14. How much money does Pete have left in the bank? _____

   Fill in how many of each coin you have.

   Nickels   _____

   Dimes   _____

   Quarters   _____

Money – Grades 2/3

Student-Centered Lesson

In this lesson, the activities themselves facilitate student learning by providing real opportunities to earn, save, and spend money. Decision making is also supported, as students must decide what to use their money to purchase and when to actually spend it. The use of a children's story at the beginning of the lesson allows students to watch as the main character goes through many of the same actions and decision-making moments that they themselves will be experiencing.

Lesson Outcome	Students will be able to add and subtract different amounts of money; students will prioritize wants, determine if and when it may be necessary to earn and save money, make reasoned purchases, and explain their reasoning.
NCTM Standards	Number and Operations:
	Understand numbers, ways of representing numbers, relationships among numbers, and number systems.
	Understand meanings of operations and how they relate to one another.
	Compute fluently and make reasonable estimates.
	Data Analysis and Probability Standard:
	Formulate questions that can be addressed with data and collect, organize, and display relevant data to answer them.
	Communication Standard:
	Organize and consolidate their mathematical thinking through communication.
	Communicate their mathematical thinking coherently and clearly to peers, teachers, and others.
	Analyze and evaluate the mathematical thinking and strategies of others.
	Use the language of mathematics to express mathematical ideas precisely.

	Connections Standard:
	Recognize and apply mathematics in contexts outside of mathematics.
Materials Needed	Black/whiteboard or overhead; coins (pennies, nickels, dimes, quarters), enough for each student to have the equivalent of a dollar and to "pay" for jobs completed; items for sale in the class store; the book *You Can't Buy a Dinosaur with a Dime* by Harriet Ziefert; student worksheets:
	8.11: Money Grab Bag
	8.12: How Much Is That Dinosaur?
	8.13: Money Earned, Money Spent
Time Required	Accessing prior knowledge – 45 minutes; reading aloud – 30 minutes; class-store application activity – 30 minutes a day, over the course of a week
Teaching Practices	Reading aloud, individual or targeted, small-group instruction, student inquiry, authentic assessment, student reflection
Assessment Purposes	Diagnostic – Accessing prior knowledge
	Formative – Class discussions, one-on-one with teacher, if desired; worksheets 8.11, 8.12, and 8.13 (during the course of the learning)
	Summative – Worksheet 8.13, plus final computation

Teaching Guide

1. Pose the questions, "How much money is enough? How do you know?" and give students time to think about and discuss them.

Accessing Prior Knowledge

2. Distribute *Student Worksheet 8.11: Money Grab Bag.*

   a. Set up bags with several pennies, dimes, nickels, and quarters.

   b. Three or four students work with each bag.

c. Students take turns picking two coins out of the bag.

d. Using their worksheets, the students draw the coins selected, label them with their names and their values inside or underneath, add their total value, and return the coins to the bag.

e. Other students in the group check each player's work and help when needed.

f. Students complete at least three rounds.

g. Students answer the self-assessment questions and return worksheets to the teacher.

Students needing more targeted intervention may benefit from work with picture representations of coins—or with real ones. Create addition problems with the pictures or coins, and write the real values underneath them. This will allow students to see the problem as a basic computation (such as 10 + 10 = 20).

3. Give students collections of pennies, nickels, dimes, and quarters. Review the names of the coins and their values by carrying the initial questions over and asking additional questions, like these:

- "Where is a dime? If you have a dime, how much money do you have? If you want to buy something for a quarter, do you have enough? How do you know?"

- "If you need fifteen cents and you have a nickel and five pennies, is that enough? How do you know?"

- "Show me which coin will be enough to buy something that costs 25 cents. What is that coin called?"

For students having difficulty correctly naming or associating value with coins, this would be a good time to review or reteach, individually or in small groups. If this is the case, students needing more targeted intervention may benefit from work with picture representations of coins—or with real ones.

It also may be necessary to do a mini-lesson on adding and subtracting money. Create addition problems with the pictures or coins, and write the real values underneath them. This will allow students to see the problem as a basic computation (such as 10 + 10 = 20). With pictures that are less realistic, like the circles with coin amounts inside, the appearance of the problem is slightly different, but the basic computation process is the same.

4. Read the story *You Can't Buy a Dinosaur with a Dime* by Harriet Ziefert.

   a. Ask comprehension questions to be sure students are following the story as well as Pete's computations and decision-making process.

   b. Students complete *Students Worksheet 8.12: How Much is that Dinosaur?* as documentation of their comprehension of the story, the computation involved, and their thinking.

   c. Students compare responses in small groups, and then groups share with the class as a whole.

Using What We Have Learned

1. Students are given bags of coins (pennies, nickels, dimes, quarters) equaling $1.00 and are introduced to the classroom store, where they can shop as long as they have the money to do so.

2. They also learn about ways that they can earn money doing chores that will help the classroom to run more smoothly—like emptying the trash ($.02 per full basket) or organizing different areas of the room ($.02 per organized area), collecting or filing papers ($.05 per class set filed, $.01 for collecting or handing out), serving as class librarian ($.10 per day), and so on. Openings are posted and can be applied for daily. Jobs are awarded based on need, behavior, and qualifications.

3. Once or twice per day, students have the opportunity to visit the class store and make purchases. Initially, the teacher should be the shopkeeper; depending on how long the store is kept running, this could become a job that pays a higher "salary" because of the mathematical qualifications necessary.

4. Distribute *Student Worksheet 8.13: Money Earned, Money Spent.*

   a. Students use this worksheet as an interactive log, keeping records of jobs that they have held, what they learned, and what they earned from them; recording purchases and amounts spent; and documenting rationale or reasons for buying.

   b. Calculations can be verified by checking against actual coins left.

   c. Logs are submitted daily for review and feedback, by either the teacher or a classmate.

5. At the end of the week, students calculate the amount of money that they should have left.

   a. They count their coins and compare their calculations to reality.

b. Students should check any discrepancies by going over their sheets of daily calculations.

6. Final calculations as well as Worksheet 8.13 can be collected and kept as evidence of student learning and thinking.

Worksheet 8.11
Money Grab Bag

Round 1

Round 2

Round 3

Name _____   Date _____

Money Grab Bag *(Cont'd.)*

Self-Assessment

1. How well can I name coins?

   _____

   _____

   Did I always name them correctly? _____

2. How well do I know what each of the coins is worth?

   _____

   _____

   Did I always know the coin's value? _____

3. How well can I add the values of coins?

   _____

   _____

   Was my addition always right? _____

4. What might I need help with when we learn more about money?

   _____

   _____

5. What could I help other people be able to understand about money?

   _____

   _____

Name _____   Date _____

How Much Is That Dinosaur?

1. What made Pete decide to buy the dinosaur?

_____

_____

_____

_____

2. How much did the dinosaur cost? Show the missing prices:

   Price of dinosaur =   _____

   Sales tax = + _____

   Total cost =   _____

3. How much money did Pete have left in his bank after buying the dinosaur?

_____

4. How does Pete feel about this?

_____

_____

_____

_____

How Much Is That Dinosaur? *(Cont'd.)*

5. What does Pete decide to do to earn more money?

_____

_____

_____

6. How many coins did Pete earn for his allowance?

Nickels _____

Dimes _____

Quarters _____

What is the total value of Pete's allowance? _____

7. From his chores and allowance, how much money did Pete earn?

Chores = _____

Allowance = + _____

Total = _____

8. What does Pete decide to sell? Why?

_____

_____

9. Would you have done the same thing? Why or why not?

_____

_____

How Much Is That Dinosaur? *(Cont'd.)*

10. Pete buys another dinosaur. Including tax, how much did Pete spend?

    Dinosaur = _____

    Tax = _____

    Total = _____

11. Why did Pete decide to buy another dinosaur?

    _____

    _____

    _____

12. How much money does Pete have left in the bank? _____

13. What would you be willing to spend all of your money to buy?

    _____

    _____

    _____

    _____

Name _____ Date _____

Money Earned, Money Spent

Employment

Date	Job	Why I applied	What I learned	Money earned
			Total earned	

End-of-Week Reflection

- Which job was the most interesting? Why do you think that?

- Which job paid the most money?

- Which job would you most like to do again? Why?

Money Earned, Money Spent *(Cont'd.)*

Purchases

Date	Purchase	Reason bought	Cost
		Total spent	

End-of-Week Reflection

- Which of your purchases was the wisest? Why?

- Which of your purchases was the most foolish or the least practical?

- Which of your purchases is your favorite? Why?

Money – Grades 4/5

Teacher-Directed Lesson

In this lesson, the teacher directs student learning by modeling. Student independent work is limited to completing a series of numeric problems.

Lesson Outcome	Students will be able to add and subtract different amounts of money.
NCTM Standards	Number and Operations:
	Compute fluently and make reasonable estimates.
Materials Needed	Blackboard or overhead, student worksheets:
	8.14: Independent Practice
	8.15: Word Problems
Time Required	1 hour
Teaching Practices	Modeling, small-group or individual instruction, direct instruction
Assessment Purposes	Formative – Scenarios; Worksheets 8.14 and 8.15
	Summative – Worksheets 8.14 and 8.15, if used for grading

Teaching Guide

1. Ask students how many of them get an allowance, and when they get it, whether they save the money or spend it. Explain that each time they save the money, they are adding amounts. If they spend the money they are subtracting.

2. Begin by having students focus on adding money. Model, on the black/whiteboard or overhead, using the following scenarios.

Scenario 1

Let's say that you earn $10.00 per week for your allowance. You decide to save it for two weeks in a row. How much money do you have after two weeks?

$$\begin{array}{r} \$10.00 \\ +10.00 \\ \hline \$20.00 \end{array}$$

The teacher points out to the students that it is important (1) to have decimal points lined up to keep place values in order and (2) to include the dollar symbol ($) as a label.

Scenario 2

Let's say that you earn your allowance for one week ($10.00) and get some extra money by earning $3.50 for every A on your report card. You earned two As.

Two As	Total for the week
$3.50 +3.50 $7.00	$10.00 + 7.00 $17.00

Explain to students what has to happen in order to correctly add the two $3.50, and again point out the need to make sure that decimal points are aligned and the labels are $.

3. Distribute *Student Worksheet 8.14: Independent Practice.*

4. Provide correct answers for students to self-check (other options for correcting these answers: have selected students come to the board or overhead and show their work, then correct it as a class; or the teacher can collect the problems and correct or grade them).

Shift students to a review of basic subtraction with money. Model, using Scenarios 3 and 4.

Scenario 3

Let's say that you have $23.00 to spend at the mall. You want to buy a CD that costs $14.00, including tax. How much money will you have left?

$$\begin{array}{r} \$23.00 \\ -14.00 \\ \hline \$9.00 \end{array}$$

Again remind students of the importance of lining up the decimal points to preserve the place value.

Scenario 4

Consider that you have $50 and need to buy three items at the store. The items total $48.39. How much change will you have left?

$$
\begin{array}{r}
\$50.00 \\
-48.39 \\
\hline
\$1.61
\end{array}
$$

Point out that it is important to make sure to regroup when subtracting.

5. Distribute *Student Worksheet 8.15: Word Problems*. Students work independently on the word problems.

6. When all students are finished, the teacher gives correct answers and collects student work to grade.

7. The teacher concludes the lesson by saying, "You will have many opportunities in your life to earn and spend money. Always remember to keep your place value by aligning your decimal points."

Independent Practice

Complete the following problems:

$2.50
+1.25

$7.25
+9.10

$12.45
+10.40

$23.99
+51.12

$41.40
+2.98

Worksheet 8.15

Word Problems

Work independently on the following word problems. Hand them in when you are finished.

What's the difference between $4.02 and $3.75?

How much change will you get if you spend $15.00 of $20.00?

How many $0.25 bags of potato chips can you buy with $2.00?

Money – Grades 4/5

Student-Centered Lesson

This lesson is a simulation involving students in a birthday shopping expedition. Working within a pre-established budget, students use store flyers to "purchase" items. They track their spending and make decisions about additional purchases based on the money that remains. Differentiation occurs through the use of manipulatives and coupons.

Lesson Outcome	Students will be able to add and subtract using money; understand the concept of budget and the skill of staying within one; make predictions; explain choices and evaluate their own and others' work.
NCTM Standards	Number and Operations Standard:
	Compute fluently and make reasonable estimates.
	Communication Standard
	Organize and consolidate their mathematical thinking through communication.
	Connections Standard:
	Recognize and apply mathematics in contexts outside of mathematics.
Materials Needed	Diagnostic worksheet with related advertisements or flyers from a local store, copies of Toys"R"Us sale flyers for each student, coupons and coins for differentiation, reflective prompts, student worksheets:
	8.16: What Do I Know?
	8.17: Shopping List
	8.18: Toys"R"Us Shopping Trip
	8.19: Shopping-Trip Checklist
Time Required	4 hours 30 minutes
Teaching Practices	Accessing prior knowledge; diagnostic, formative, and summative assessments; student centered; peer feedback; criteria in the form of a checklist; student reflection; performance-based assessment
Assessment Purposes	Diagnostic – Worksheet 8.16
	Formative – Worksheets 18.17, 18.18 (during lesson), and 18.19 (if used for feedback and revision during the lesson)
	Summative – Worksheets 18.18 and 18.19

Teaching Guide

Accessing Prior Knowledge

1. Pose several questions to students:

   - Why do stores print advertisements or sales flyers?

   - How can people use advertisements or sales flyers to get more for their money?

   - Why do people care about getting a "good buy"?

   - What does it mean to have a budget?

   - If you wanted to find the best buy on something, how would you know where to shop?

   - In addition to advertisements and sales flyers, what else do stores do to help people save money and make them want to shop there instead of somewhere else?

2. Share with students that they are going to go on an imaginary shopping trip and will have to stay within a budget. They will have flyers from stores to help them determine what they can buy and how much. It will be important for them to keep track of their spending so that they stop when they run out of money.

Lesson Diagnostic

3. Distribute *Student Worksheet 8.16: What Do I Know?* and begin the actual lesson with a diagnostic activity.

   a. Students are instructed to "spend" $25.00 on school supplies, using an advertisement from a local store.

   b. Using the worksheet provided, students keep track of what they are buying and how much each thing costs, and they also maintain a running account of the money spent and how much money is left.

4. Collect students' work and assess it to determine what, if anything, needs to be taught, reviewed, or retaught before proceeding with the rest of the lesson.

Lesson Activity – Going Shopping

5. At the beginning of this part of the lesson, students discover that their imaginary shopping trip will be to Toys"R"Us, using a budget of $85.00—money that they received as birthday presents from friends and relatives.

6. Distribute *Student Worksheet 8.17: Shopping List*.

   a. Individually, students brainstorm a list of the things they think they would most like to buy.

   b. Each student prioritizes his or her list by numbering the items in order of importance.

   c. Students draw a line underneath the item that they predict will be the last one that they can afford on their $85.00 budget.

7. Distribute *Student Worksheet 8.18: Toys"R"Us Shopping Trip* and Toys"R"Us sale flyers.

   a. Students are instructed to use their flyer and worksheet to determine and document their purchases and computations, and to explain decisions they make.

   b. They shop from their prioritized list but are free to make changes to their list or the order of the items, based on their shopping experience.

   c. They can also make changes to their purchases by returning any item and adding its cost back into the remaining total.

8. Check student work for understanding of the necessary computation and process. Based on how students are progressing, differentiate at this time, as follows:

   a. Students who appear to be having difficulty or who are becoming frustrated can be given play money to use to "pay" and "make change," documenting each amount spent and the total remaining, then checking with a calculator and adjusting. Each calculator adjustment should be noted. Students could include in their reflections or explanations how using the calculator to check their work affected how much they were able to purchase.

   b. Students for whom this activity seems appropriate should be told that they have an additional $25.00 to spend. These students add this to their current total and continue to shop.

   c. Students who are more advanced in their understanding of money and ability to compute with it can use real or teacher-made coupons (dollar amounts off, "two-for," percent off, or a combination, depending on the ability level and experience of the students) and apply them, taking advantage of additional sales when they make their "purchases." Each coupon used should be documented and its impact on their ability to purchase explained. This adds a layer of complexity to the activity that will keep these learners engaged and challenged.

9. Distribute *Student Worksheet 8.19: Shopping-Trip Checklist.*

   a. Using a checklist, students work with a partner to give and receive feedback on their shopping.

   b. They revise their work based on this feedback.

10. When they have finished shopping, students answer the following reflective questions:

   • How did you know you were done?

   • What changes, if any, did you make to your list? Why?

   • How close was your prediction about the items you could afford to buy? Why do you think this was?

   • How did you use the feedback that your partner gave you? How did your feedback help your partner?

   • What have you learned about adding and subtracting money?

At any time during the Going Shopping lesson activity (steps 5 through 9), formative assessment is possible if student work is reviewed in progress, either in brief conferences with or between students, or by checking student worksheets as they are being used. Feedback can be given orally or in writing, or by using a copy of the checklist. This provides an opportunity to intervene with those students who might need content or process help or refocusing, while allowing those students who are progressing to continue to work independently.

Name _____ Date _____

What Do I Know?

Use the advertisement to help you decide how to spend $25.00 on school supplies.

1. Keep track of what you're buying and how much each thing costs, as well as a running account of the money spent and how much money is left.

Item	Cost	Money spent	Money left
		$0.00	$10.00
Subtotal			
Subtotal			
Subtotal			
Subtotal			
Subtotal			
Subtotal			
TOTAL			

2. At the end of this part of the lesson, the teacher will collect your work, and assess it to determine what, if anything, needs to be taught before proceeding with the rest of the lesson.

Shopping List

Things I would most like to buy at Toys"R"Us:

1. Prioritize your list by numbering the items in order of importance.

2. Draw a line underneath the item that you predict will be the last one that you can afford on your $85.00 budget.

Name _____   Date _____

Toys"R"Us Shopping Trip

Item bought or returned	Cost	Money spent	Money left	Reason for purchase or return
		$0.00	$85.00	
Subtotal				
Subtotal				
Subtotal				
Subtotal				
Subtotal				
Subtotal				
Subtotal				
Subtotal				
Subtotal				
TOTAL				

Name _____ Date _____

Shopping-Trip Checklist

☐ It is easy to read the names of the items I am buying.
Suggestions for improvement:

☐ The cost of each item is clearly written to show dollars and cents.
Suggestions for improvement:

☐ The cost of each item is labeled.
Suggestions for improvement:

☐ The reason for each item I bought or returned is clear.
Suggestions for improvement:

☐ I stayed within my budget.
Suggestions for improvement:

☐ Computations are accurate.
Suggestions for improvement:

☐ My work is neat and easy to follow.
Suggestions for improvement:

Chapter 9

Fractions

Fractions are more than just numerators and denominators. They are representations of reality. Life rarely presents itself in neatly tied-up packages of whole and even amounts. Houses are sold with "2.5 bathrooms," a recipe requires 1/4 of a teaspoon of salt, shoes sizes increase by halves, and information travels in nanoseconds. All involve fractions of a whole. Understanding the relationship between parts and the whole that they compose is an important concept for students to grasp and be able to use, in school and in life.

The lessons in this chapter are designed to provide a range of approaches to the topic of fractions, based on student age and also on prior experience. Teacher-directed examples rely primarily on a combination of direct instruction and guided practice, as well as independent work. They are straightforward and specific, and they fit well inside a limited amount of time. Student-centered lessons rely more on inquiry and exploration, engaging students in actively constructing an understanding of fractions and their meaning and importance. If desired, parts of teacher-directed lessons can be used in combination with the student-centered lessons, as introductions or mini-lessons, or for targeted small-group instruction as needed.

Fractions – Grades K/1

Teacher-Directed Lesson

This lesson uses sets of objects, equal parts of sets, or both as a way of introducing young students to the concepts of whole and part, focusing on students' abilities to recognize and create representations of equal parts of a whole. It addresses the question, "What is half?" by having the teacher present examples and lead students through activities designed to help them comprehend and create "half."

Lesson Outcome	Students will be able to understand that a whole can be made up of parts and also to identify half of a whole when dealing with concrete objects or familiar ideas.
NCTM Standards	Number and Operations Standard: Understand numbers, ways of representing numbers, relationships among numbers, and number systems.
Materials Needed	Counters (or other manipulatives); chart paper, overhead, or black/whiteboard; Student Worksheet 9.1: How Many?
Time Required	1 hour—could be broken into three 20-minute segments
	Questions 1–5, whole class
	Question 6 pairs
	Question 7 individual
Teaching Practices	Explanation, modeling, small-group instruction (optional as needed)
Assessment Purposes	Formative – Responses during 1–5, responses to 6c
	Summative – Worksheet 9.1

Teaching Guide

1. Show students a whole apple. Then cut it in half.

    a. Ask students to tell what they saw you do.

    b. Explain that you cut the apple in half; show students that each half is the same size, and that when you fit them back together, you have a whole apple.

2. Repeat this with another familiar object, like a sandwich or a candy bar.

3. Explain that parts can make a whole thing, or parts can make a collection.

4. Create a group of four students, two boys and two girls.

    a. Ask the class to count how many students are in the group. When they count four, write the number 4 on a piece of paper and post it above the group, repeating that there are four students in the group.

    b. Ask how many of the four students are girls. After getting the correct response, write the number 2 on a piece of paper and give it to the two girls to hold, stating that two of the four students are girls.

    c. Repeat this, asking the class how many of the four students are boys. Give the two boys the piece of paper with 2 written on it, saying this time that two of the four students are boys.

    d. Summarize for students that there are four in the group: two boys and two girls. Restate this by saying that we could also say that half of the group is boys and half are girls.

5. Repeat this with a group of six or eight students, half boys and half girls, ending again with a summary and statement that links the same number in each group to being half of the whole group.

6. Show students a drawing of six squares, three that are blue and three that are red.

    a. Ask students to count how many squares they see. When they count six, ask how many of the six squares are blue. Draw a ring around the three blue squares.

    b. Ask how many squares are red; draw a ring around them.

    c. This time, instead of summarizing, ask students how they could say what they see. Prompt them to elicit both the number description (three of the squares are red, three of the squares are blue) and the fraction description (half of the squares are red, half of the squares are blue).

At this time, it may be appropriate to work with a small group of students who seem to need more guided practice. In this case, the teacher can provide small-group instruction on step 6 with this group, walking around the room only for the responses to 6c.

7. Divide students into pairs and give each pair a group of ten manipulatives (popsicle sticks, counters, cubes, and the like), five of one

color and five of another. Tell students that the collections they have are half color A and half color B.

a. Have students count the total number of items in the group and draw what they see, including colors.

b. Have students separate their collection by color, count each color group, and draw what they see.

c. Walk to each pair and ask the students to show you half.

Students who have difficulty showing half may benefit from a reteaching, one on one or in a small-group setting. It is recommended that this reteaching occur prior to giving these students Worksheet 9.1 or that the worksheet become the basis of the reteaching.

8. Distribute one copy of *Student Worksheet 9.1: How Many?* to each student. Read the worksheet for those who may have difficulty reading, but have each student complete the activities individually. Collect this from students as evidence of where they are in their understanding.

Name _____ Date _____

How Many?

1. Draw a ring around the picture that shows half of a square colored black.

2. Count the stars in the group.

   a. How many stars are there? _____

   b. How many of the stars are black? _____

   c. How many white stars do you count? _____

   d. Explain how much of the group of stars is black.

_____

_____

   e. Explain how much of the group of stars is white.

_____

_____

How Many? *(Cont'd.)*

3. Draw six circles in the space.

a. Color one half of the circles blue.

b. How many circles are blue? _____

Fractions – Grades K/1

Student-Centered Lesson

This lesson uses inquiry-based explorations as a strategy for introducing young students to the concepts of whole and part, focusing on students' abilities to recognize, create, and discuss various representations of equal parts of a whole. Each exploration is designed to connect to students' experiences as they engage with the question, "What is half?"

Lesson Outcome	Students will understand the concept of whole and also identify parts within a set and parts of a whole, specifically one half, when dealing with concrete objects or familiar ideas and solving real-life problems. Students will explain their thinking and reflect on their own learning, as well as raise new questions.
NCTM Standards	Number and Operations Standard:
	Understand numbers, ways of representing numbers, relationships among numbers, and number systems.
	Problem-Solving Standard:
	Build new mathematical knowledge through problem solving.
	Monitor and reflect on the process of mathematical problem solving.
	Reasoning and Proof Standard:
	Make and investigate mathematical conjectures.
	Communication Standard:
	Organize and consolidate their mathematical thinking through communication.
	Communicate their mathematical thinking coherently and clearly to peers, teachers, and others.
	Use the language of mathematics to express mathematical ideas precisely.

Materials Needed	Chart paper; apples; pretzel sticks; marsh-mallows; precut shapes (hearts, circles, squares and equilateral triangle); clay; other manipulatives and student worksheets:
	9.2: Testing Our Hunches
	9.3: What Is Half?
Time Required	2.5 hours total, including 20 minutes per exploration, 30 minutes for testing hunches, and 30 minutes for individual, end-of-lesson assessment plus ongoing processing and reflection time
Teaching Practices	Questioning, small-group and whole-class work, student inquiry, individual or small-group targeted instruction (optional as needed)
Assessment Purposes	Diagnostic – Student responses to initial conversation about "What is half?"
	Formative – Student Worksheet 9.2, as well as class discussions after student reflection in each exploration and end-of-class reflection
	Summative – Student Worksheet 9.3

Teaching Guide

1. Have a conversation about halves; ask for examples from real life, beginning with the example of asking for half a cup of coffee or drinking half a glass of milk. What other examples of half can students share?

2. Once you have several familiar examples of half, ask students the question, "What is half?" Record their responses on a piece of chart paper.

3. Explain to students that they are going to try to discover more about what half is and then, in a little while, they'll come back to the list to see if there is anything that they want to change or add.

The following explorations should be completed sequentially, but if time or attention is a concern, they can be scheduled at different times over the course of one or two days. In this case, it becomes important for "hunches" to be posted, visible, and revisited before a new exploration is begun.

Explorations can be modified to accommodate varying levels of independence by structuring the pairs or groups to support independent work or to allow the teacher to participate. Pairs and groups can remain consistent through all explorations, or they can be changed to better accommodate the learning needs that become apparent. Teacher participation in a group may consist of asking more or simpler questions, scaffolding the work of the group, or providing a presence to increase student focus and engagement.

Exploration 1: How Big Is Half?

1. Cut apples in half, but try not to separate them completely, so they can still be seen as whole. There should be enough apples so that every student will get a half.

2. Assign a pair of students to each apple.

3. Ask students to draw the apple that they see.

4. Have students each take half of the apple. Remind them that they have half, and ask them what they notice about the size of the halves.

5. Reflection: Is each half the same size as the other?

6. In small groups of three or four, students discuss this question. Groups report out their thinking to the class as a whole.

7. Based on ideas generated and the exploration questions, "How big is half?" the teacher or the students or both should state an "exploration hunch."

   Exploration 1 hunch: Halves are the same size as each other *or* halves are equal.

   It may be possible or interesting to refine this hunch by asking students if *all* halves are equal to all other halves. To approach this, have students compare the half of the apple that they have to a half from a different pair by looking at size and shape and even trying to fit them together to make a "whole." Have students explain what they think about this and how it fits with their hunch.

8. Write and post the exploration hunch.

Exploration 2: How Many Halves Make a Whole?

1. Give each student a heart-shaped piece of paper with a dotted line down the center.

2. Have students fold their hearts in half along the dotted line.

3. Prompt students to each look carefully at the heart, both when it is open and when it is folded in half.

4. Reflection: What makes half different from whole? In small groups of three or four, students discuss this question. Groups report their thinking to the class as a whole.

5. Based on ideas generated and the exploration question, "How many halves make a whole?" the teacher, the students, or both should state an "exploration hunch."

    Exploration 2 hunch: There are two halves in a whole.

6. Write and post the exploration hunch.

Exploration 3 – How Much Is Half?

1. Each pair of students gets a set of eight items (such as pretzel sticks, pencils, stickers, or teddy-bear counters) that they must share equally.

2. Prompt students to count all of the items. Write "Total: 8" on a piece of chart paper.

3. Tell students to share the items so they each have half. If they need help deciding what they are supposed to do, remind them of the other exploration hunches. Solicit from the class their ideas of how to go about dividing the items. Students are likely to show multiple strategies in their solutions: some may count to determine half of the group, some may break each group member piece in half. Prompt students to explore the fact that both solutions are possible, though one or the other may be more appropriate depending on the situation; for example, splitting a group of people.

4. Reflection: How do I know I have shared half? In small groups of three or four, students discuss this question. Groups report their thinking to the class as a whole.

5. Based on ideas generated and the exploration question, "How much is half?" the teacher and/or the students should state an "exploration hunch."

   Exploration 3 hunch: Each half is the same amount, or each half is an equal amount.

6. Write and post the exploration hunch.

Testing What We've Discovered – How Good Are Our Hunches?

Students should work in small groups of three or four. Distribute one copy of *Student Worksheet 9.2: Testing Our Hunches* to each group. The results of the group's work are documented on this sheet. Students can share the responsibility of recording by changing recorders for each test, or one recorder can document all results.

Test 1: The Marshmallow Test

1. Provide each group with ten marshmallows (for highly capable students, a more challenging test would include an odd number of marshmallows).

2. Ask students to use one or more of the exploration hunches to help them show half of the marshmallows.

3. They should document their choices and a picture of the results on Worksheet 9.2.

Test 2: The Clay Test

1. Provide each group with a piece of clay and a popsicle stick "knife".

2. Instruct groups to roll their clay into a rope and use one or more of the exploration hunches to help them show half of the rope.

3. Their choices and a picture of the results should be documented on Worksheet 9.2.

Test 3: The Shape Test

1. Provide each group with a precut circle, square, and equilateral triangle (see the following example).

2. Have groups use one or more of the exploration hunches to help them color half of each shape (for the square, they can color it in as many ways as possible).

3. Their choices and a picture of the results should be documented on Worksheet 9.2.

Individual Assessment

1. Distribute *Student Worksheet 9.3: What Is Half?* (Students who are not independent writers may need assistance documenting their thinking and writing explanations.)

2. End-of-lesson class reflection: Return to the original list of responses to the question "What is half?" and ask students to share what they now know about *half* that is different from what is on the original sheet.

3. At this point, ask students what, if anything, should be changed on their first list so that it matches what they know now.

Date _____

Testing Our Hunches

Names of group members: _____

Test	Picture of half	Hunches that helped us
The Marshmallow Test		
The Clay Test		
The Shape Test		

Which hunches were most helpful? _____

_____

_____

Were any of our hunches wrong? _____

Do any of our hunches need to be updated? _____

Name _____   Date _____

Worksheet 9.3

What Is Half?

Use the hunch hints to help you remember what we have learned about half.

1. Draw a ring around the picture that shows half of a square colored black.

[Hunch hint: Think about what we learned from looking at our hearts.]

2. Count the stars in the group.

How many stars are there? _____

How many of the stars are black? _____

How many white stars do you count? _____

Which of our hunches do these stars show?

_____

_____

_____

What Is Half? *(Cont'd.)*

3. Draw six circles in the following space.

Color one half of the circles blue.

How many circles are blue? _____

Why would you say that half of the circles are blue?

_____

_____

_____

Which hunches did you use to help you answer this question?

_____

_____

_____

Fractions – Grades 2/3

Teacher-Directed Lesson

This lesson is designed to help students begin to think of, and talk about, parts of a whole or part of a group as fractions. The teacher uses visual and numeric representations to help students consider the fractions 1/2, 1/3, 1/4, 1/6, and 1/10.

Lesson Outcome	Students will be able to recognize and illustrate the fractions 1/2, 1/3, 1/4, 1/6, and 1/10 as they relate to a specific part of a whole.
NCTM Standards	Number and Operations Standard: Understand numbers, ways of representing numbers, relationships among numbers, and number systems.
Materials Needed	Chart paper, overhead, or white/blackboard; red and blue crayons or colored pencils; Student Worksheet 9.4: Understanding Fractions
Time Required	45 minutes
Teaching Practices	Explanation, modeling, guided practice, check for understanding
Assessment Purposes	Formative – Responses to class work and student responses to Student Worksheet 9.4, question 1 Summative – Student Worksheet 9.4, question 2

Teaching Guide

1. On overhead, chart paper, or black/whiteboard, show students these representations:

One whole
1

One half
1/2

One quarter
1/4

2. Distribute *Student Worksheet 9.4: Understanding Fractions* and have students complete question 1.

At this point, the teacher should check student responses for understanding. If reteaching or further review is necessary, this should take place before moving to the next part of this lesson.

3. Return to the images used for the initial review at the beginning of the lesson. Show students that the total number of segments of a figure is the same as the bottom number of the fraction, and the number of shaded sections is the same as the top number of the fraction. If students do not already know them, this is an opportunity to introduce the mathematical terms *numerator* and *denominator*. If they are familiar with the terms, then they should use them instead of "top number" and "bottom number" in the explanation.

4. On an overhead or chart paper, use fraction bars to represent the fractions 1/3, 1/6, 1/10.

One third
1/3

5. Explain that each segment is 1/3 of the whole bar, and show how, if you shade two of the segments, you have colored two thirds of the bar.

One sixth
1/6

6. Explain that each segment is 1/6 of the whole bar. Shade three more segments of the bar, to show what 4/6 of the bar looks like.

One
tenth
1/10

7. Explain that each segment is 1/10 of the whole bar, and have students tell you how many more to shade to make 3/10 (6/10, 8/10, 10/10). Once you have shaded 10/10, remind students that when the numerator and denominator are the same, you have one whole.

8. Have students complete the fractions bars in their student worksheets. Collect these as evidence of student understanding.

Name _____   Date _____

Understanding Fractions

1. Shade each of the figures so that it represents the fraction.

One quarter   One whole   One half
1/4   1   1/2

2. Shade 1/3 of this bar. Then shade another 1/3.

How much of the bar is shaded?

3. Shade 2/4 of this bar.

How much of the bar is *not* shaded?

4. Shade 1/6 of this bar blue. Then shade 3/6 more of the bar red.

How much of the bar is shaded?

5. Shade 7/10 of this bar.

6. How many more tenths would you have to shade if you wanted to show one whole?

Fractions – Grades 2/3

Student-Centered Lesson

This lesson is designed to help students think and talk about fractions as a way of describing parts of a collection. In the context of analyzing survey data, students use visual and numeric representations to understand and apply the fractions sixths and tenths.

Lesson Outcome	Students will organize survey data and use fractions to analyze and describe the data and tell its story.
NCTM Standards	Number and Operations Standard:
	Understand numbers, ways of representing numbers, relationships among numbers, and number systems.
	Communication Standard:
	Organize and consolidate their mathematical thinking through communication.
	Communicate their mathematical thinking coherently and clearly to peers, teachers, and others.
	Use the language of mathematics to express mathematical ideas precisely.
	Representation Standard:
	Create and use representations to organize, record, and communicate mathematical ideas.
	Use representations to model and interpret physical, social, and mathematical phenomena.
Materials Needed	Chart paper, overhead, or white/blackboard; student worksheets:
	9.5: And the Survey Says . . .
	9.6: Telling a Data Story

Time Required	3 hours, total. Can be broken up into segments as follows:
	60 minutes (guided inquiry: 1–18)
	30 minutes (independent practice, Worksheet 9.5)
	30 minutes storytelling (small group)
	60 minutes (independent practice)
Teaching Practices	Questioning; student inquiry; guided and independent practice; small-group work; individual or small-group instruction (as needed), or both
Assessment Purposes	Formative – Responses to class work, Student Worksheet 9.5
	Summative – Student Worksheet 9.6, student reflection

Teaching Guide

1. Select ten students in the class to participate in a survey about their favorite color. Use tally marks to show the results on chart paper; for example:

   Red – ///

   Blue – ////

   Yellow – /

   Green – //

   Explain that this is survey data, and that data tells a story, but sometimes it's hard to understand the story because the data isn't organized or shown in a way that is easy to read.

2. Ask students to help you organize the data. What could be done to the tallies to make them easier to read? What could be done to the order of the colors to help us see which colors are the most and the least popular? For example, the new arrangement of the preceding data might be:

   Blue – 4

   Red – 3

   Green – 2

   Yellow – 1

Now ask students what helps them to read a story and stay interested. When someone mentions pictures, explain that data stories need pictures too.

3. Use an adaptation of a fraction bar to depict the results, coloring the bar with the appropriate colors. Using the preceding results, the bar would look like this:

blue	blue	blue	blue	red	red	red	green	green	yellow

4. Explain that now their survey data is organized and has a picture to go with it.

5. With the students, tell the story that the organized data shows. Write the story on an overhead or chart paper, so that it is visible. For example: "Ten students were surveyed about their favorite color. Five of the ten surveyed liked blue. Three of the ten liked red best. Two of the ten students said that green was their favorite color, and only one of the ten students liked yellow the most." (Some students may look at the picture and say that most of the students liked blue, or they may make comparative statements like "more liked red than yellow." Teachers can prompt these kinds of statements to push students' thinking, but the basic information in the story is all that is needed to proceed with the lesson.)

6. Point out that one out of ten students said that yellow was their favorite color, and put that on its own fraction bar:

yellow									

7. Write the fraction 1/10 underneath the yellow-colored bar.

8. Point out that three out of ten students said that red was their favorite color and put that on its own fraction bar:

red	red	red							

9. Write the fraction 3/10 underneath the red-colored bar.

10. Ask students what they notice when they look at the fraction and the bar. They can write their observations down, but do not take their responses yet.

11. Post the bar that shows the students who chose blue:

blue	blue	blue	blue						

12. Write the fraction 4/10 underneath.
13. Have students work in groups of three or four to share what they see. Let groups report and document what they say.
14. Post the last bar, showing green responses:

green	green								

15. This time, ask groups to decide what fraction belongs underneath, based on what they have observed. They should write the fraction on a piece of paper and hold it up so everyone can see.
16. Groups should then report their thinking to the class as they state the fraction. If, at any time, a group hears something that makes the members want to change their response, they can do so, but then when it's their turn to share their thinking they must include both what they were thinking when they wrote their original answer and their reason for changing their response.
17. At this point, the teacher and students should sum up what the class has discovered: the total number of sections is the number on the bottom of a fraction and the number of colored sections is the top number. This is a good opportunity to introduce the mathematical terms *numerator* and *denominator*, if students are not already familiar with them. If these terms are already part of the language that students are using, then they should replace "number on the bottom" and "top number" in their summary.
18. Have students revise their data story to include fractions. (For example, "Ten students were surveyed about their favorite color. 4/10 surveyed liked blue. 3/10 liked red best. 2/10 said that green was their favorite color, and only 1/10 liked yellow the most.")
19. Distribute *Student Worksheet 9.5: And the Survey Says . . .* Have students work individually to complete the activities. Then allow them to work in pairs or triads to create their data story.
20. Groups share data stories. The teacher can ask questions like, "What do our stories have in common?" "How are our stories different?" "What makes a good data story?" This last question can lead to a checklist that students can use when they complete Worksheet 9.6.

Fractions

21. Collect this as evidence of student learning and use the information to help determine which students, if any, need assistance, review, or reteaching before moving on.

22. Distribute *Student Worksheet 9.6: Telling a Data Story.*

23. Reflection:

    • How do fractions help to communicate data and tell a data story that's easy to understand?

    • What does a denominator mean with an apple? With a group of people?

Name _____ Date _____

And the Survey Says . . .

Look at the survey data organized and illustrated in the following bars, and answer the questions that follow.

Number of Students

	9			
	8			
	7			
	6			
	5			
	4			
	3			
	2			
	1			

Chocolate   Vanilla   Strawberry

Favorite Flavor of Ice Cream

1. What was the total number of students who were surveyed?

2. Out of all of the students surveyed, how many said that vanilla was their favorite flavor?

   Answer: _____ out of _____ said that vanilla was their favorite flavor of ice cream.

3. What fraction of the students who were surveyed said that vanilla was their favorite ice cream flavor? _____

4. Was there any flavor that was liked best by 1/10 of the students who were surveyed?

5. How many students preferred chocolate? How would you write this as a fraction?

   _____

6. How many liked strawberry best? How would you write this as a fraction? _____

7. Working with one or two of your classmates, write the story that this survey data tells you. Be prepared to share your story with the rest of the class.

Name _____   Date _____

Telling a Data Story

On this worksheet you will find data from a school survey. Once you have organized and illustrated the data, use it to help you tell the story, "How We Get to and From School."

Survey 1 – How do we get to and from school?

Survey data:

    Walk – //

    School bus – ///

    Other (car, train, city bus) – /

1. Organize the survey data in the table. Remember to change tallies to numbers and think about an order that helps to make the information clear.

Survey data with numbers
  Walk –
 	
  School bus –
 	
  Other (car, train, city bus) –
	Survey data in order

2. What was the total number of students surveyed about how they travel to and from school? _____

3. Illustrate this data by shading the correct number of boxes for each.

Number of Students

7 6 5 4 3 2 1

School Bus   Walk   Other

Transportation to and from school

Telling a Data Story *(Cont'd.)*

What fraction would you use to tell how many of the students walk to and from school?

_____

How would you write the number of students who take the bus to and from school as a fraction? _____

What fraction would tell how many of the students surveyed travel to and from school in

some other way? _____

Fractions – Grades 4/5

Teacher-Directed Lesson

The following lesson focuses on developing students' conceptual understanding of fractions and of their relative size. In this lesson, the teacher controls both the content and the process, as learning opportunities are crafted so that the teacher leads the class through the activities. At the end of the lesson, students are assigned problems, which they complete independently and check when the teacher reads the answers.

Because the content of this lesson involves comparing fractions with different denominators, it is most appropriate for students who already know about fractions and are able to add or subtract fractions with like denominators. This lesson will support students as they begin to learn about ordering a series of fractions, from least to greatest or greatest to least.

Lesson Outcome	Students will be able to compare fractions with different denominators.
NCTM Standards	Number and Operations:
	Understand numbers, ways of representing numbers, relationships among numbers, and number systems.
Materials Needed	Overhead, chart paper, or black/whiteboard; student notebooks; Student Worksheet 9.7: Which Is Greater?
Time Required	45 minutes
Teaching Practices	Explanation, modeling, review, direct instruction, guided practice, independent practice
Assessment Purposes	Formative – Student Worksheet 9.7, 1–5
	Summative – Student Worksheet 9.7

Teaching Guide

1. Review the vocabulary *numerator* and *denominator*, asking questions to check for understanding. For example:
   - Looking at the fraction 1/3, who can tell me what the numerator is?
   - Who can identify the denominator in 4/5?
   - What's the name of the 8 in the fraction 6/8?

2. Tell students that when you compare fractions, the rule is, "The bigger the denominator, the smaller the amount." Have students write this rule in their notebooks.

3. Using an overhead projector, black/whiteboard, or chart paper, show students the differences among a variety of fractions. Use a combination of numbers (1/2, 1/3, 1/4, 1/5, 1/6, and so on) and pictures illustrating the fractions to show how each time the denominator gets bigger, the actual amount gets smaller.

4. Give students *Student Worksheet 9.7: Which Is Greater?* and have them complete comparisons 1–5.

5. Review responses, giving or eliciting correct answers.

6. Assign the rest of the worksheet as needed, for class work or homework.

At this point, it may be appropriate to meet with a small group of students who are struggling, while the rest of the class works independently. It is also possible to pair students or save the rest of the worksheet for reinforcement or review.

Name _____   Date _____

Which Is Greater?

Part 1

Assume that all fractions refer to the same whole. Circle the larger fraction:

1. 1/4   3/4

2. 2/3   2/6

3. 1/5   1/2

4. 6/7   6/8

5. 1/9   1/5

6. 4/7   2/7

7. 5/8   3/9

8. 1/3   1/2

9. 3/7   3/9

10. 2/5   4/5

Which Is Greater? *(Cont'd.)*

Part 2

Assume that all of the fractions are of a whole group of twenty-four people. Circle the largest.

1. 1/6   1/4

2. 7/8   2/6

3. 2/3   1/2

4. 3/4   3/8

5. 1/3   1/6

6. 2/3   2/4

7. 5/8   3/6

8. 1/3   1/2

9. 3/8   3/4

10. 2/6   4/6

Part 3

Explain a situation in which 1/4 would be bigger than 3/4 (hint: you will need more than one whole). When you are finished, hand your paper in.

Fractions – Grades 4/5

Student-Centered Lesson

In the following lesson, students develop a conceptual understanding of fractions and their relative size using an inquiry-based approach. They engage directly with problems and, as a result, discover and draft their own rules for the comparison of fractions. The lesson begins with a diagnostic assessment to determine what students already know about the connection between the size of a fraction and its denominator. This provides the teacher with assessment data that will allow the teacher to recognize and address the individual needs of students as the lesson progresses. The ideas generated during the diagnostic assessment are also posted, and become a pool from which students can select in order to test their rules and that can be used to stimulate the generation of additional ideas for testing. By the end of the lesson, each student will have defined a rule or rules for determining the size of a fraction based on its denominator.

Lesson Outcome	The student will generate a valid rule for comparing fractions and will use that rule to solve problems involving relative size (such as which is greater?).
NCTM Standards	Number and Operations Standard:
	Understand numbers, ways of representing numbers, relationships among numbers, and number systems.
	Reasoning and Proof Standard:
	Make and investigate mathematical conjectures.
	Communication Standard:
	Communicate their mathematical thinking coherently and clearly to peers, teachers, and others.
Materials Needed	Chart paper, overhead, or black/whiteboard; student worksheets:
	9.8: What Do I Already Know About the Size of Fractions?
	9.9: Discovering and Testing Rules
	9.10: Reflecting on What I Have Learned

Time Required	3 hours
Teaching Practices	Diagnostic assessment; facilitating student inquiry (establishing, testing, and revising a rule), incorporating student reflection
Assessment Purposes	Diagnostic – Worksheet 9.8, 1–4
	Formative – Worksheet 9.9, 1–3 (scheduled conferences, checking student work, and so on)
	Summative – Worksheet 9.10, revised rule and its application to question 8, correct explanation of how rule was used

Teaching Guide

1. Give students *Student Worksheet 9.8: What Do I Already Know About the Size of Fractions?*

2. Accessing prior knowledge: Introduce the lesson by asking what students already know about the connection between the size of a fraction and its denominator. Give students five minutes to think about the question and write down any ideas they may have.

3. Ask students to share their responses orally, and post these on chart paper, the overhead, or the black/whiteboard, without correcting or commenting on them. Keep this list public and advise students that during the unit they can either test these ideas or come up with others and test those. Inform them that by the end of the lesson they will have decided on a rule or rules for determining the size of a fraction based on its denominator.

4. Ask students to complete exercise 2 in Worksheet 9.8 and to explain their answers.

   If students have difficulty with this exercise, it may evidence a gap in understanding about fractions. If so, some intervention may benefit students at this point, before continuing with further activities in the lesson. Depending on how many students exhibit difficulties, the teacher may decide to engage them in small-group targeted instruction (refocusing or reteaching as needed); or students could work with a buddy or in small groups; in which the focus is to uncover misunderstandings or confusions, try to resolve difficulties using the expertise in the group, and then share the remaining questions or concerns with the whole class for their help and the help of the teacher.

5. Proceed to question 3 in Worksheet 9.8. Students who have difficulty with this question may have issues with following sequenced directions, or may need content intervention related to understanding fractions of a whole or the fraction vocabulary (such as *fourth* and *third*).

6. Number 4 in Worksheet 9.8 can provide diagnostic information about an individual student's ability to understand fractions as parts of a whole, to understand the concept of dividing a whole into parts, to make visual comparisons, and to write an explanation of a concrete activity.

7. At this point, it is appropriate to check students' work (questions 1–4 in Worksheet 9.8). This could be at the end of the first day or at another break during the lesson. Assessing the students' success in completing these questions will determine their readiness to complete the remaining activities. Some students may need intervention before proceeding. This might be a good time to use all or part of the teacher-directed fraction lesson for grades 4/5 (with a small group or as a homework review.

8. Introduce question 1 in Worksheet 9.9, if necessary, by repeating the questions on the worksheet.

9. Introduce questions 2–4 in Worksheet 9.9, prompting students as necessary.

Differentiating Process

Not all processes work for all learners. Sometimes there is a conflict between student learning style and pedagogy or teaching style. This can interfere with learning or can make it difficult to accurately and reliably assess what has been learned. For an activity like the birthday-cake problem in worksheet 9.9, question 7 which relies heavily on linguistic abilities, paper folding is a strategy that will support students who are visual or kinesthetic learners, enabling them to physically manipulate their solution and see the answer.

Here's how to use paper folding in the birthday-cake problem:

1. Give each child a blank sheet of paper.

2. Explain that they will be treating the paper as though it were the birthday cake.

3. For each step in the problem that involves a fraction, students fold the paper:

   You are told that 2/3 of the cake is for you and your nine relatives:

   - Fold the paper into thirds.
   - Label two of those thirds "nine family members and me."
   - Fold the two thirds into ten equal sections, one each for you and nine relatives.

Your four friends can share the other 1/3 of the cake:

- Label the last third "four friends."
- Fold that third into four equal sections.

Who will get a bigger piece of cake, a friend or a relative?

Looking at the pieces and comparing their sizes, students will see that their friends got bigger pieces of birthday cake.

4. Conclude the lesson with the reflective activity of Worksheet 9.10. Have students again look at the original list of ideas that they presented earlier and then answer the following reflective questions in writing, rephrasing them if necessary:

- How close were we?
- Now, what can you tell me about the denominator of a fraction and its size?
- What do you understand now that you didn't know when we first made this list?

Students can be given the opportunity to share their responses to any or all of these questions as a way of wrapping up the lesson.

Name _____ Date _____

Worksheet 9.8

What Do I Already Know About the Size of Fractions?

1. What can you tell about how the denominator of a fraction is related to its size? You have five minutes to write down some ideas.

My Ideas:

What Do I Already Know About the Size of Fractions? *(Cont'd.)*

2.
☐ ☐ ☐

☐ ☐ ☐

a. Color in 1/2 of the squares.

b. Circle 1/3 of the squares.

Which is more, 1/2 or 1/3? Explain your answer.

What Do I Already Know About the Size of Fractions? *(Cont'd.)*

3.

```
┌─────────────────────────────────────────────────┐
│                                                 │
└─────────────────────────────────────────────────┘
```

   a. First, divide the bar in half.

   b. Next, divide one of the halves into fourths.

   c. Then divide the other half of the bar into thirds.

```
┌─────────────────────────────────────────────────┐
│ Which segments are bigger, the thirds or the    │
│ fourths? Why do you suppose that is?            │
│                                                 │
│                                                 │
│                                                 │
│                                                 │
│                                                 │
│                                                 │
│                                                 │
│                                                 │
│                                                 │
│                                                 │
│                                                 │
│                                                 │
└─────────────────────────────────────────────────┘
```

What Do I Already Know About the Size of Fractions? *(Cont'd.)*

4.

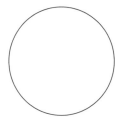 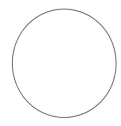

Here are two pizzas. Divide one so that it can be evenly shared among eight friends. Divide the other so that it can be shared among five friends.

Who gets the bigger piece of pizza, the friends who each get 1/8 or the friends who each get 1/5? Why is that?

Name _____   Date _____

Worksheet 9.9

Discovering and Testing Rules

1. Look at the list of ideas about the size of a fraction and its denominator that we shared earlier. Do any of them seem true, based on what we've just done?
   a. Use one of the ideas that you believe is true, or another idea that you have, to help you to write a rule about the size of a fraction and its denominator.
   b. Check your rule to make sure that it is true for all three examples that we have just completed. Write it in the box.

Draft rule:

Name _____ Date _____

Discovering and Testing
Rules *(Cont'd.)*

2. Create another example that proves your rule. Share it with a classmate. Can he or she follow the example? Does your rule work?

Example:

Discovering and Testing Rules *(Cont'd.)*

3. Use your rule to help you respond to five of your classmates' examples. Keep track of how well your rule works. If you find that you want to revise your rule along the way, feel free to do so.

Student name	My rule worked, and this is why:	My rule didn't work, and I am thinking I should try . . .

4. What have you learned about your rule?

Notes about what I've learned:

Discovering and Testing
Rules *(Cont'd.)*

5. Revise your rule so that it is more accurate or easier to understand. Share it with at least two other students and make more changes if necessary.

Revised rule:

6. Use your rule to help you decide which is bigger. Circle your responses:

    a. 1/6 or 1/9

    b. 1/4 or 1/3

    c. 1/8 or 1/2

    d. 1/7 or 1/5

How did your rule help you to decide?

7. Use your rule and the space on this page to help you solve the following story problem:

It's your birthday, and your family is celebrating by having a cake. Nine people from your family are with you to celebrate, and so are four of your friends.

It's time to cut the cake. You are told that 2/3 of the cake is for you and your nine relatives. Your four friends can share the other 1/3 of the cake.

    a. Who will get a bigger piece of cake, a friend or a relative?

    b. How do you know you are right?

Reflecting on What
I Have Learned

Look at the original list of ideas that we posed. Respond to the following questions in writing:

1. How close were we?

2. Now, what can you tell me about the denominator of a fraction and its size?

3. What do you understand now that you didn't know when we first made this list?

Resources and Web Links

Balanced Assessment in Mathematics Project

From 1993 to 2003, the Balanced Assessment in Mathematics Program operated at the Harvard Graduate School of Education. The project group developed a large collection of innovative mathematics assessment tasks for grades K–12, and trained teachers to use these assessments in their classrooms.

The library of over three hundred mathematics assessment tasks developed during the project remains freely available through this Web-site. Teachers may use these materials in their own classrooms at no cost.

www.balancedassessment/concord.org

Center for Excellence for Science and Math Education

This site contains classroom ideas for math and science teachers.

http://nsm.fullerton.edu/cesme.html

The Change Site (Eisenhower National Clearing House – ENC)

Resources to facilitate discussion and reflection on improving K–12 mathematics.

www.goenc.com

Exemplary and Promising Mathematics Programs

Descriptions of ten programs (from Connected Math to Middle-School Mathematics through Applications Project), selected by an expert panel convened by the U.S. Department of Education.

www.enc.org/external/ed/exemplary

Videopaper by the Math Forum's Bridging Research and Practice Group. Includes video clips that illustrate the group's discussion and suggestions for effective math interventions and strategies.

http://mathforum.org/teachers/

Jobs for Kids Who Like Math

The Bureau of Labor Statistics helps kids understand what it takes to be an engineer, accountant, mechanic, cashier, architect, or computer scientist.

http: Stats.bls.gov/k12/html/edu_math.htm

Mandalas

Download patterns for mandalas.

www.junemoon.com/free.html

www.mariposamuseum.org/downloads/mandala_sample.gif

www.Math.com

Lesson plans, career resources, homework help, puzzles, and quizzes.

Math Archives

This site contains links to over 150 great sites for educators to get classroom ideas.

http://archives.math.utk.edu/k12.html

Mathematics: What's the Big Idea?

Eight workshops covering patterns and functions, data, geometry, whole numbers, ratio and proportion, algebra, and the future of math—with lesson plans and a lot of links.

www.learner.org/resources/series98.html

Math Forum

A large, searchable collection of excellent math resources, from math news to teaching resources. Home of "Ask Dr. Math," an online math question and answer service for students. The Math Forum @ Drexel, 3210 Cherry Street, Philadelphia, PA 19104.

www.forum.swarthmore.edu

www.mathforum.org

Math in Daily Life: How Do Numbers Affect Everyday Decisions?

"Playing to Win," "Cooking by the Numbers," "Population Growth," and more are in this eight-part series. Lesson plans and links on subjects ranging from probability to geometry.

www.learner.org/interactives/dailymath/

Math in the Real World

Dozens of adult professionals—from business executives to computer programmers—provide practical answers to the question, "Why should I study math?" From the Mathematical Association of America, which also sponsors an annual Math Awareness Month and looks at math's connections to fields from manufacturing to medicine.

www.maa.org/careers/index.html

Mosaics

Downloadable mosaic patterns.

www.hobbycraft.co.uk/LESSONS/springmosaic_2.htm

The National Council of Teachers of Mathematics

This site contains some of the latest ideas that are helping the nation's math classrooms.

www.nctm.org

Quilts

Downloadable quilt patterns.

www.quilt.com/ColoringBook/QuiltColoringBook.html

The Teaching Gap: Best Ideas from the World's Teachers for Improving Education in the Classroom

Excerpts from an excellent book by Jim Stigler and James Hiebert, based on the Third International Mathematics and Science Study (TIMSS) research that stimulated development of The Missing Link series. Ideas, advice and links to other TIMSS-related sites.

www.lessonlab.com/teaching-gap/index.htm

TIMSS Resource Center

An excellent starting place for learning more about TIMSS, which inspired The Missing Link series. With links to video clips of classroom practice, detailed research findings, and helpful materials for teachers and curriculum specialists.

www2.rbs.org/ec.nsf/pages/L2TIMSS

References

Brownell, W. (1946). The measurement of understanding, prepared by the society's committee. In N. B. Henry (Ed.), *National Society for the Study of Education: Committee on the Measurement of Understanding*. Chicago: University of Chicago Press.

Chappuis, S., Stiggins, R. J., Arter, J., & Chappuis, J. (2004). *Assessment for learning: An action guide for school leaders*. Portland, OR: Assessment Training Institute.

Cobb, P., Wood, T., Yackel, E., Nicholls, J., Wheatley, G., Trigatti, B., & Perlwitz, M. (1991). Assessment of problem-centered second-grade mathematics project. *Journal for Research in Mathematics Education, 22,* 3–29.

Cotton, K. (2001, October). *Classroom questioning, close-up 5*. School Improvement Research Series, Northwest Regional Educational Laboratory.

Dyer, M. K., & Moynihan, C. (2000). *Open-ended questions in elementary mathematics: Instruction and assessment*. Larchmont, NY: Eye On Education.

Ernst, P. (1988, August). *The impact of beliefs on the teaching of mathematics*. Paper presented at the Sixth International Congress of Mathematical Education, Budapest, Hungary.

Gall, M. (1984). Synthesis of research on teacher's questioning. *Educational Leadership, 42,* 40–46.

Gall, M. D., Ward, B. A., Berliner, D. C., Cahen, L. S., Winne, P. H., Elashoff, J. D., & Stanton, G. C. (1978). Effect of questioning techniques and recitation in student learning. *American Educational Research Journal, 15:* 175–199.

Martin-Kniep, G. (2001). *Becoming a better teacher: Eight innovations that work*. Alexandria, VA: ASCD.

National Research Council. (2002). *Helping children learn mathematics*. Mathematics Learning Study Committee, J. Kilpatrick, & J. Swafford, Eds. Center for Education, Division of Behavioral and Social Sciences in Education. Washington, DC: National Academy Press.

Samson, G. E., Strykowski, B., Weinstein, T., & Wahlberg, H. J. (1987). The effect of teacher questioning on student achievement. *Journal of Educational Research, 80:* 290–295.

Sanders, S.M. (1966). *Classroom Questions: What kinds?* New York: Harper & Row.

Sitko, M. C., & Slemon, A. L. (1982). Developing teachers' questioning skills: The efficacy of delayed feedback. *Canadian Journal of Education, 7,* 109–121.

Swift, J. N., & Gooding, C. R. (1983). Interactions of wait-time feedback on questioning instruction on middle school science teaching. *Journal of Research in Science Teaching, 20,* 721–730.

Walsh, J. A., & Sattes, B. D. (2003). *Questioning and understanding to improve learning and thinking: Teacher manual* (2nd ed.). Charleston, WV: Appalachian Regional Educational Laboratory.

Walsh, J. A., & Sattes, B. D. (2005). *Quality questioning: Research-based practice to engage every learner*. Thousand Oaks, CA: Corwin Press.

Index